Irrigated Forage Production

Developments in Crop Science

Developments in Crop Science 24

Irrigated Forage Production

by

AMOS DOVRAT

Faculty of Agriculture,
The Hebrew University of Jerusalem,
Rehovot, Israel

In collaboration with:

Yehezkel Cohen and **Amos Goldman**

Agricultural Research Organization,
Bet Dagan, Israel

ELSEVIER

Amsterdam – London – New York – Tokyo 1993

ELSEVIER SCIENCE PUBLISHERS B.V.
Sara Burgerhartstraat 25
P.O. Box 211, 1000 AE Amsterdam, The Netherlands

ISBN 0444-88300-2

This book is printed on acid-free paper.

Printed in The Netherlands

And Ahab said unto Obadiah, Go into the land, unto all fountains of water, and unto all brooks; peradventure we may find grass to save the horses and mules alive, that we lose not all the beasts.
I Kings: 18–5

PREFACE

The subject of this monograph is the improvement of forage production for livestock in areas with limited resources of irrigation water and sometimes of land. At present many irrigated forage production systems are associated with low water use efficiency, wasteful utilization practices and considerable losses of nutrients during conservation. Often, the abundance of water available for irrigation did not evoke the need of specific attention to the amount, timing and method of water application, nor to the selection and management of forages in order to produce maximum nutrient yields. However, in today's increased competition, with alternative production options and shrinking water resources, the use of irrigation water for forage production creates an increasing challenge in maintaining this activity economically viable.

In developed countries the improvement of forage quality in general and of irrigated forage in particular, has lately received considerable attention in order to meet the demand for high-quality animal products. In less developed countries there is an increasing demand for animal products because of the steep rise in the size of the population, which in specific areas requires the development of efficient irrigated forage production systems.

This monograph represents an attempt to collate and discuss within one volume the many aspects of forage production when crops are grown under irrigation. The review relies heavily on recently published research, progress reports and growers' experience. The reader is constantly referred to additional sources of information, specifically on subjects to which only limited attention could be paid in the context of this book. The material is subdivided into two parts: the first provides the reader with some basic elements of forage yield formation, and the second part provides details on five crops plus pasture selected because of their wide distribution, or because they are recommended for use. The main topics covered in this book deal with crop agronomy, e.g. soil management, crop selection, input scheduling, and forage quality; and with physical processes such as climate factors, crop-water functions, low water quality, and irrigation methods.

I am greatly indebted to Drs. Y. Cohen and A. Goldman of the Agricultural Research Organization, for contributing Chapters 3 (Water requirements) and 11 (Economics), respectively. Parts of the book have been read in draft form by Drs. A. Arieli, I. Arnon, D.W. Grimes, U. Kafkafi, T. Kipnis, Y. Leshem, K.F. Lowe, M.J. Pinthus, W.A. Williams and D. Yaron, and the authors are very grateful for their constructive comments. Vivian Priel edited an early version of the manuscript; Ruthy Felzman did the secretarial and typing work; and Niha Gestetner drew the diagrams. Last but not least, the librarians of the Central Agricultural Library at Rehovot and of the AFRC Institute for Grassland and Animal Production at Hurley rendered useful services in the retrieval of pertinent references. The manuscript of this book could not have been submitted to the publisher without the efficient and kind assistance of each of the above mentioned and many other persons.

Special thanks to my wife Tamar and our family, for their understanding, cooperation and loving care.

Amos Dovrat

TERMINOLOGY AND SYMBOLS USED

ADF	Acid detergent fiber
B	Boron
BMR	Brown-midrib mutant
C_3	Calvin cycle of CO_2 fixation
C_4	Hatch–Slack cycle of CO_2 fixation
Ca	Calcium
CGR	Crop growth rate
CF	Crude fiber
Cl	Chlorine
CO_2	Carbon dioxide
CP	Crude protein
CWC	Cell wall content
d	Day
DE	Digestible energy
DM	Dry matter
DMD	Dry matter digestibility
DOMD	Digestible organic matter in dry matter
dS	Decisiemen
EC	Electrical conductivity
E_p	Pan evaporation
ET	Evapotranspiration
Fe	Iron
g	Gram
ha	Hectare
IVDM	*In vitro* digestible matter
IVOMD	*In vitro* digestible organic matter
K	Potassium
K_c	Crop coefficient
kg	Kilogram
kPa	Kilopascal
LAI	Leaf area index
m	Meter
ME	Metabolizable energy
Mg	Megagram, ton
Mg	Magnesium
MJ	Megajoule
Mn	Manganese

Mo	Molybdenum
MPa	Megapascal
N	Nitrogen
NDF	Neutral detergent fiber
NH_4	Ammonium
NIR	Near-infrared radiation
$N\text{-}NH_4$	Nitrogen as ammonium
$N\text{-}NO_3$	Nitrogen as nitrate
nm	Nanometer
ODM	Organic dry matter
P	Phosphorus
P_n	Net photosynthesis
R	Respiration
RGR	Relative growth rate
S	Sulphur
T	Transpiration
TNC	Total nonstructural carbohydrates
TDN	Total digestible nutrients
TRC	Transpiration coefficient
WUE	Water use efficiency
Zn	Zinc

CONTENTS

Chapter 1

BACKGROUND

Forage is defined as food from plants or plant parts eaten by domestic animals, which is either gathered *in situ* (pasture) or whole-crop harvested and hauled for feeding on the farm. Forages are either fed fresh (green chop) or conserved and, because of their perishable nature, made into hay, haylage, silage or preserved by other methods. Each forage species has specific growth and quality characteristics which are modified by environmental conditions, including management. Feeding and management systems differ markedly, depending on local economic, environmental and social factors. For example, dairy-farm enterprises may run the gamut from feeding purchased forages and concentrates to zero-grazed, intensively penned cows, to feeding cattle on grazed pasture and conserved pasture throughout the year (Scott, 1982).

The choice between management options in the growing and utilization of irrigated forage crops often focuses on the: (i) degree of water use efficiency; (ii) availability of relatively many potential forage crops; (iii) effect of harvesting method and timing on forage quality; (iv) structure of the forage and animal production system; (v) costs of purchased roughages and concentrates; and (vi) specific nutritional requirements of different classes of animals in space and in time. Alternatively, producing irrigated forages for the market requires inputs similar to many other quality-defined industrial crops. The above list of management parameters is by no means exhaustive; however, it draws attention to the complexity of irrigation water–crop–animal interrelationships.

Current trends in dairy-farm enterprises emphasize:

(i) increase in herd size (Quesenberry, 1985; Teese, 1992);

(ii) limiting the number of different forages fed in order to maintain a uniform forage supply;

(iii) maximizing intakes of high quality forages (Van Soest, 1982);

(iv) automation of mixing and feeding equipment in the preparation of complete rations (Quesenberry, 1985; Shefet, 1990);

(v) new technologies in forage conservation and mechanization of harvesting (Bolsen, 1985; Larsen and Rider, 1985); and

(vi) integrating forage crops with pasture (Pritchard et al., 1989).

Identification of optimum forage strategies is needed to face these new challenges. Comprehending the physiological and agronomic potential of the crops reviewed in this book may assist in making more efficient use of forage and water resources.

REFERENCES

Bolsen, K.K. 1985. New technology in forage conservation-feeding systems. 15th Int. Grassl. Congr. (Kyoto, Japan), pp. 82–86.

Larsen W.E. and Rider, A.R. 1985, Mechanization of forage harvesting and handling. pp. 452–459. In: M.E. Heath, R.F. Barnes and D.S. Metcalfe (Eds.) Forages. Iowa State Univ. Press, Ames, IA.

Pritchard, K.E, Mason, W.K., Stockdale, C.R. and Moran, J.B. 1989. Integrating forage maize with pasture legumes for efficient dairy production in southeastern Australia, 16th Int. Grassl. Congr. (Nice, France), pp. 1143–1144.

Quesenberry, N. 1985. Future of the California dairy industry to the year 2000. 15th Calif. Alfalfa Symp. (Fresno, CA), pp. 92–103.

Scott, J.D.J. 1982. Efficiency of dairying under contrasting feeding and management systems in North America, Israel, Europe and New Zealand. 14th Int. Grassl. Congr. (Lexington, KY), pp. 843–846.

Shefet, G. 1990. The feeding system for ruminants in Israel. Hassadeh Quart. 1: 78–80.

Teese, I. 1992. Economic forage production and utilization. pp. 86–93. In: B. Bartsch and W. Mason (Eds.), Feedbase 2000. CSIRO and Dairy Res. Devel. Corp., Melbourne, Australia.

Van Soest, P.J. 1982. Nutritional Ecology of the Ruminant. O and B Books, Corvallis, OR. 374 pp.

Part One:
The Biomass

Chapter 2

YIELD FORMATION

In contrast to most other arable crops, the harvested or economic yield of forage crops equals the biological yield. This whole-crop yield usually consists of the total above-ground plant mass, but sometimes also of the below-ground mass, as in the case of fodder beet. Yield formation is the product of many complex processes, and cannot be dealt with in depth here. Three major processes are carried out consecutively by crops: interception of solar radiation by the leaf canopy, conversion of the intercepted radiant energy to plant dry matter (DM), and partitioning of the DM produced between plant components. Reviews of the physiology of crop yield are provided by Eastin et al. (1969), Milthorpe and Moorby (1974), Evans (1975), Hay and Walker (1989), and of modelling aspects of crop growth by van Keulen and Wolf (1986). In this part of the book we will discuss some biological concepts of the accumulation of plant mass in terms of forage yield, concomitant with high digestibility, and maximum water and nutrient use efficiency.

2.1 MORPHOLOGY

The growth habit of forage plants affects light penetration into the canopy, development patterns, distribution of assimilates, regrowth after defoliation, propagation, and many other determinants of crop productivity.

The plants of most forage crops, like those of other crops, pass through two successive phenological development phases: vegetative and reproductive. During *vegetative* development all activities are associated with the generation and expansion of leaves, the formation of terminal meristems and the concurrent expansion of the root system. *Reproductive* development starts as soon as the shoot apex, whether apical or axillary, differentiates to floral structures. The phasing of phenological development in time and space often dictates the mode of utilization. Most forage crops (e.g. alfalfa, berseem clover, grasses) are cultivated for the amount of

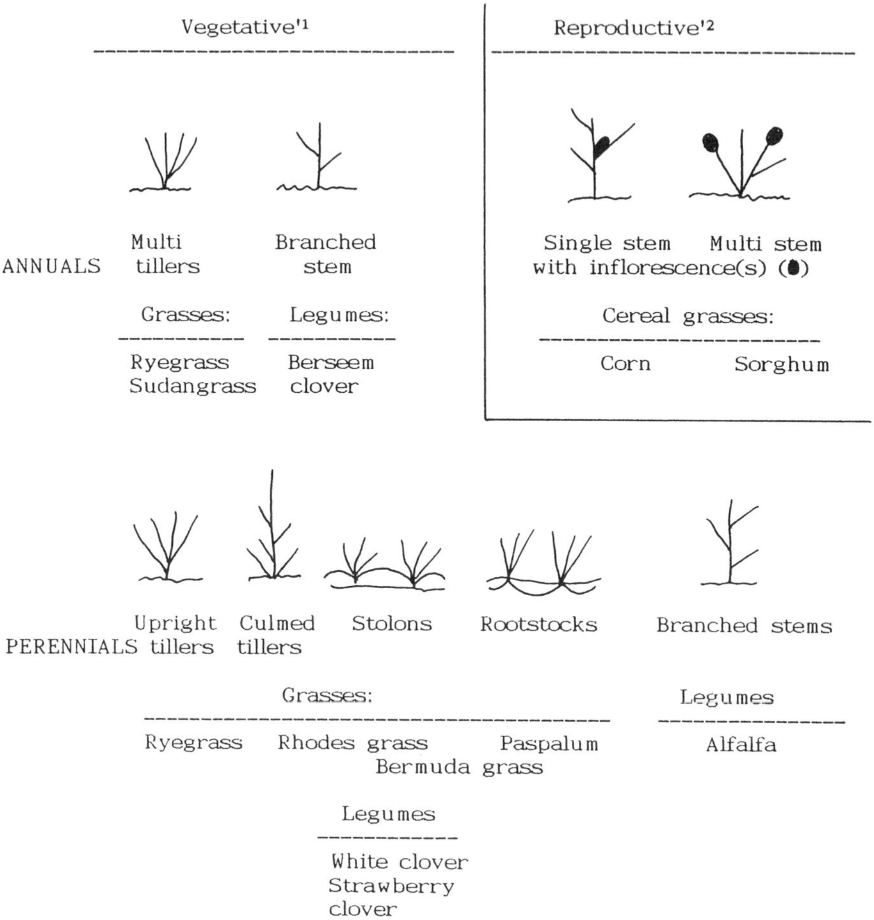

Fig. 2.1. Schematic morphology of forage plants. 1. Vegetative: most of the plant weight is derived from vegetative growth. 2. Reproductive: the reproductive fraction constitutes the major portion of total plant weight at harvest.

vegetative growth they contribute to the yield and are often harvested (or grazed) periodically. In other crops (e.g. forage corn and forage sorghum) the product of reproductive growth, i.e., kernels, is a major component of the total yield, and these crops are usually harvested once only.

Common morphogenetic types of forage plants are shown in Figure 2.1. The designations 'vegetative' and 'reproductive' are chosen in order to label the major fraction contributing to the total plant weight based on morphological characteristics. Most grasses and legumes, either annual or perennial, belong to the 'vegetative' type of growth. They develop floral structures (including seed) at a certain point in time, but their fraction of the total plant mass is usually small. The shoot of grasses is called 'tiller' and that of legumes is called 'stem', the latter usually branching. Tillers develop upright or in certain species spread above-ground as stolons or below-ground as rootstocks. The 'reproductive' type of growth is that of cereal grasses (i.e., corn and sorghum), which develop one or more upright stems (or stalks) with energy-rich inflorescences, which often constitute a major fraction of the harvested plant mass. A detailed morphology of cultivated forage grasses and legumes is presented by Hyder (1972), Allan Jones (1985), Metcalfe and Nelson (1985), and Robson et al. (1988). The morphology of forage corn and fodder beet is described briefly in Chapters 7 and 10, respectively.

2.1.1 Regrowth after Defoliation

Morphological adaptation of many grasses and legumes enables them to endure repeated partial defoliation. This regrowth potential following defoliation depends on the location of apical meristems — which are the source of new leaves — in tillers or stems. In the *vegetative* tiller, new developing leaves and sheaths, and new tillers develop from buds on the shoot apex, which is situated at or close to ground level (Fig. 2.2A). As long as tillers remain in the vegetative developmental stage, defoliation above the meristematic region of leaves may remove fully emerged laminae and any developing leaves. Regrowth depends on the growth of young leaves which are still in the process of emerging from the enclosing sheaths. Defoliation of tillers during the period of vegetative development is usually succeeded by intensive tillering.

As a result of floral induction of the shoot apex, the region of intercalary meristems of developing leaves begins to extend and carries the transformed apex upward on an elongated stem to form a terminal inflorescence. The tiller then becomes *reproductive* and the apex is likely to be removed when the plant is defoliated (Fig. 2.2B). Once the apex has been removed the tiller dies, and regrowth depends on the presence of other tillers with intact apices or on the production of new tillers. Thus, the pattern of regrowth of grass following defoliation is strongly dependent on whether the tillers are in the vegetative or reproductive state. New tillers are produced continuously during the growing season. With the onset of flower induction, the ratio between reproductive and vegetative tillers increases. Regrowth of temperate grass swards is extensively reviewed by Davies (1988).

Some subtropical grasses develop non-reproductive tillers, called *culmed vegetative tillers*, with elevated apices and thus prone to decapitation

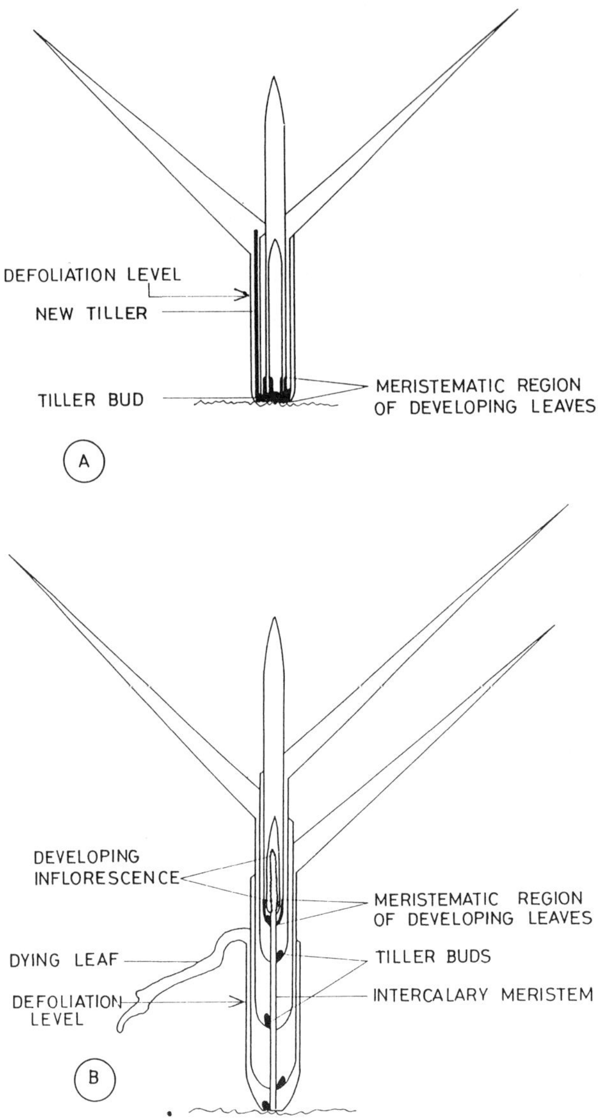

Fig. 2.2. A diagrammatic representation of a vegetative tiller (A), and a reproductive tiller (B) of perennial ryegrass. Above a given defoliation level, the vegetative tiller will regrow, whereas the reproductive tiller ceases to grow because of the removal of the meristematic region of developing leaves. (Adapted from Davies, 1973).

(Hyder, 1972). Grasses such as Rhodes grass (Kipnis, 1977; Dovrat et al., 1980), hybrid pennisetum (Warndorff et al., 1985) and others develop successive cycles of elongated tillers. Regrowth is increasingly delayed as

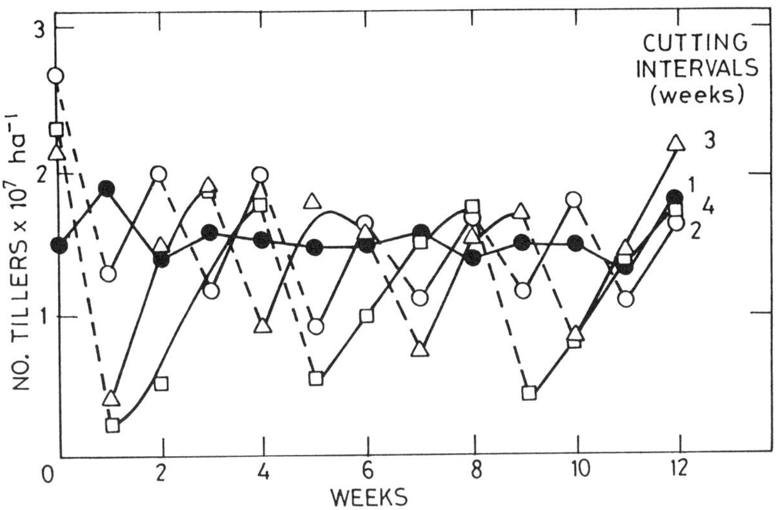

Fig. 2.3. Pattern of number of tillers following different cutting intervals of irrigated Rhodes grass cv. Katambora. Bet Dagan, Israel. (From Dayan et al., 1981).

the cutting interval is extended because fewer tillers capable of regrowth remain after cutting. An example of the fluctuations in the number of tillers as affected by the length of the cutting interval of irrigated Rhodes grass is shown in Figure 2.3. A model named TILDYN simulates the tiller dynamics and growth of Rhodes grass after defoliation (Dayan et al., 1981).

The sites of regrowth of alfalfa and berseem after partial defoliation are buds on the upper portion of the root (crown) and stems below the cutting height. Their location is shown in Figures 5.2 and 8.3.

2.2 DRY MATTER ACCUMULATION

Crops grow by accumulating matter from their environment by three processes:

(i) *uptake of water*, which constitutes approximately 70–90% of the crop mass;

(ii) *photosynthesis* (i.e. the light-dependent reduction of carbon dioxide (CO_2) from the air), which accounts for about nine-tenths of the remaining DM, and

(iii) *uptake of minerals* which account for the rest of the DM.

Water and mineral uptake occur by transfer across the soil–root interface. However, for photosynthesis only 7% enters as hydrogen in water across the root surface while 93% enters as CO_2 diffusing across the air–leaf interface. Although water accumulation is the major contribution to growth, photosynthesis (i.e., CO_2 assimilation) is quantitatively the limiting process on which accumulation of water and minerals depends. Photosynthetic transfer of CO_2 from the air to the leaves is therefore the basic rate-determining process for advanced systems of crop production (Warren Wilson, 1967).

The term 'canopy structure' encompasses the distribution and arrangement of the above-ground plant parts within a canopy, including the absolute weight and percentage contribution of the different plant parts to the community.

2.2.1 Light Interception

Shoot and root mass occupy the respective space above and below the soil surface, but surprisingly little space on a volume basis. Of the total canopy volume only 1% or less is occupied by plant tissue, whereas in terms of leaf area distribution the canopy intercepts, most of the time, more than 95% of the incoming light (Milthorpe and Moorby, 1974). All other factors being equal, DM accumulation is a function of the amount of leaf exposed to sunlight.

The amount of net radiation reaching the top of the canopy changes in a seasonal pattern according to latitude and is negatively affected by the presence of clouds or atmospheric turbidity caused by water vapor, dust, smoke and other aerosols. Energy exchange in the biosphere is reviewed by Gates (1962). Formulae for calculating average net radiation reception according to latitude and day of the year are given by de Wit (1965) and Pearson and Ison (1987). High light reception is of major importance for maximizing productivity because of the close relationship between irradiance and the net DM gain per unit leaf area (Warren Wilson, 1967). Warren Wilson also stresses the favorable radiation regime for plant growth associated with clear skies and fairly long days in arid compared with other climates. This is the major reason why arid zones have the potential to support very high crop DM production year after year, provided that water and adequate soil nutrients are available.

The subject of light interception and its measurement in crop stands is discussed comprehensively by de Wit (1965), Loomis and Williams (1969), Monteith (1969) and Woodward and Sheehy (1983), and that in pasture communities by Ludlow (1978).

Radiation

Some basic elements associated with radiation are summarized below:

(i) Solar radiation can be characterized by its wavelength (between 300 and 3000 nm) or directional components (direct beam and diffuse), and by daylength. With an overcast sky, all radiation is dispersed and reaches the crop canopy as diffuse radiation, which usually is less than 20% of that received on a very clear day.

(ii) The waveband in which radiant energy is available for photosynthesis is within 400–700 nm, corresponding to the blue and red ends of the visible spectrum. This 'visible light' called *photosynthetically active radiation* (PAR), accounts for about half of the total radiation. It is expressed as irradiance (in energy units: J m^{-2} s^{-1}).

(iii) The remaining 50% of total radiation lies in the infra-red range (wavelength >700 nm) and plays an important role in the control of plant temperature and evaporation. The ratio of PAR to total radiation changes slightly with the amount of water vapor and dust in the atmosphere but seems to be relatively insensitive to the presence of clouds.

(iv) The waveband between 700 and 730 nm, near-red radiation, is of interest because it activates phytochrome and other pigment systems that control photomorphogenetic responses (Taiz and Zeiger, 1991).

Leaf Orientation

Most of the radiation intercepted by a crop is absorbed by leaves, more specifically, by leaf laminae. Leaf sheaths, stems and inflorescences also absorb radiation and are capable of photosynthesis in most species. Some relevant geometrical (and optical) features of leaf canopies are described below:

(i) From the light falling on a leaf, roughly 80% — depending on species — is absorbed by the chloroplasts and the remainder is transmitted, reflected or absorbed by other pigments that do not contribute to photosynthesis. Thus, radiation becomes progressively reduced in energy and altered in spectral composition as it passes through a plant canopy.

(ii) The area of leaf laminae within any horizontal layer of a canopy is specified by a *leaf area index* (LAI), i.e., the area of leaves or leaf parts per unit area of ground beneath them. An example of the decline of light intensity with depth in a sugarbeet canopy with the corresponding LAI values in 5-cm horizontal layers is presented in Figure 2.4. The decrease in light reception in a canopy is modified by LAI, solar elevation, canopy structure, stage of development of the crop, and species. The foliage display and light absorption in field communities of irrigated corn at different plant densities are discussed by Loomis et al. (1968).

(iii) Leaves or leaf sections can be treated as planes making an angle with the horizontal, referred to as the leaf or foliage angle. De Wit (1965)

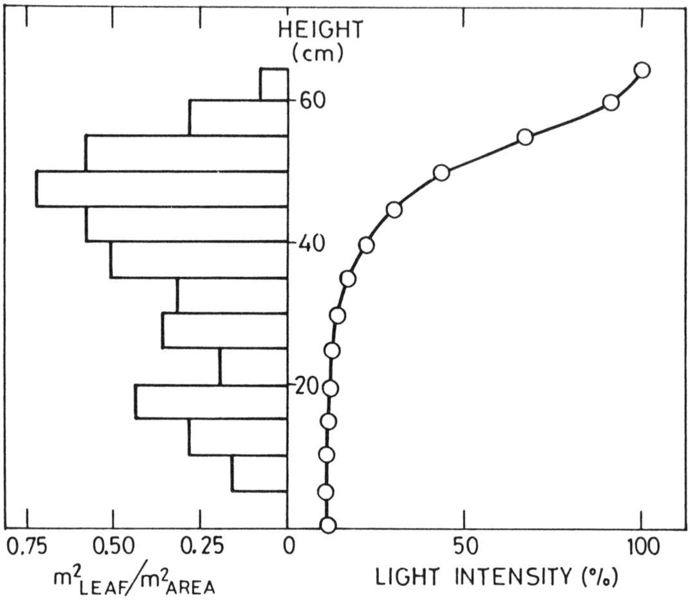

Fig. 2.4. Stratified leaf area indices and the decline of light intensity with depth in a sugar beet crop. Total LAI = 4. Light extinction is almost complete after the light penetrated approximately 30 cm into the canopy. It is assumed that a similar pattern of light distribution is found in a fodder beet crop. (From Spitters, 1986).

measured the frequency distribution of leaf angles for a number of crop species. Thus, clover had a 'planophile' (or prostrate) canopy structure with a preponderance of leaves at small angles to the horizontal. Sugarbeet was 'plagiophile' with a fairly uniform distribution of leaf angles from 0 to 90°. Most grasses, including corn and sorghum, were 'erectophile' in early growth but as the leaves grew longer and more flaccid they bent to form a planophile canopy. Erect canopies with vertical leaves usually attain higher LAI's for complete interception of light ('closed canopy') than do prostrate canopies.

(iv) In addition to LAI, leaf angles and specific crop characteristics, there are many other factors which determine the rate of light extinction with depth in a canopy, e.g. sun flecks, solar elevation, fraction of direct or diffuse radiation of total radiation, etc.

(v) The advantage in terms of final DM yield of erectophile canopies, either between cultivars of the same crop species or between crops, compared with those with more planophile canopies, is only partly substantiated by experimental evidence (Trenbath and Angus, 1975). It appears that differences in mean leaf angle are likely to be much less significant than differences in the rate at which the canopy expands to form a complete cover, in onset of leaf senescence, and in the rate of respiration per unit leaf area.

(vi) Extended green *leaf area duration* usually increases whole-crop yield.

2.2.2 CO_2 Assimilation

The rate of CO_2 exchange (net (or apparent) photosynthesis, P_n) is the difference between the rates of CO_2 uptake (gross, (or true) photosynthesis, P_g) and the emission of CO_2 (respiration, R, which is the sum of growth and maintenance respiration):

$$P_n = P_g - R$$

These processes ultimately determine the efficiency with which intercepted PAR (of which perhaps 85% is absorbed) affects the conversion of CO_2 into crop DM. The detailed biochemistry of these processes is provided by Milthorpe and Moorby (1974) and Hay and Walker (1989).

An example of the rate of CO_2 assimilation and concurrent DM production of field-grown non-stressed white clover is presented in Figure 2.5. Assimilation and respiration of this crop vary with the number of leaf layers, i.e. the higher the LAI, the more complete the light interception. At LAI = 3, in the example given, practically all the light was intercepted and consequently the contribution to P_g of leaf layers above 3 was negligible. As expected, the growth R was proportional to the amount of assimilates formed, which is represented by the curve of P_g. Conversely, the maintenance R was proportional to the biomass to be maintained and thus increased with the age of the stand. This increase in maintenance R explains the decrease in P_n when LAI exceeded 3. In this white clover crop, 40–50% of the assimilates was emitted through respiration, so that P_n was about half of P_g (Spitters, 1986).

The efficiency with which incident light is absorbed by leaves and used for CO_2 reduction to carbohydrate by photosynthesis is rather low. The actual photochemical efficiency is usually taken as being equivalent to 2–5 g DM per MJ of intercepted radiation, or 25–63 g DM m^{-2} d^{-1} where incoming radiation is 350 MJ m^{-2} d^{-1} (Pearson and Ison, 1987). The energy costs of synthesis vary according to the compounds which are produced, being greater in legumes than in grasses.

Two broad categories of species are recognized, differing in their relationship between irradiance and leaf P_n: (i) C_3 species — having the Calvin pathway of CO_2 fixation, which include temperate grasses and legumes, tropical legumes and most broadleaved plants; and (ii) C_4 species — having

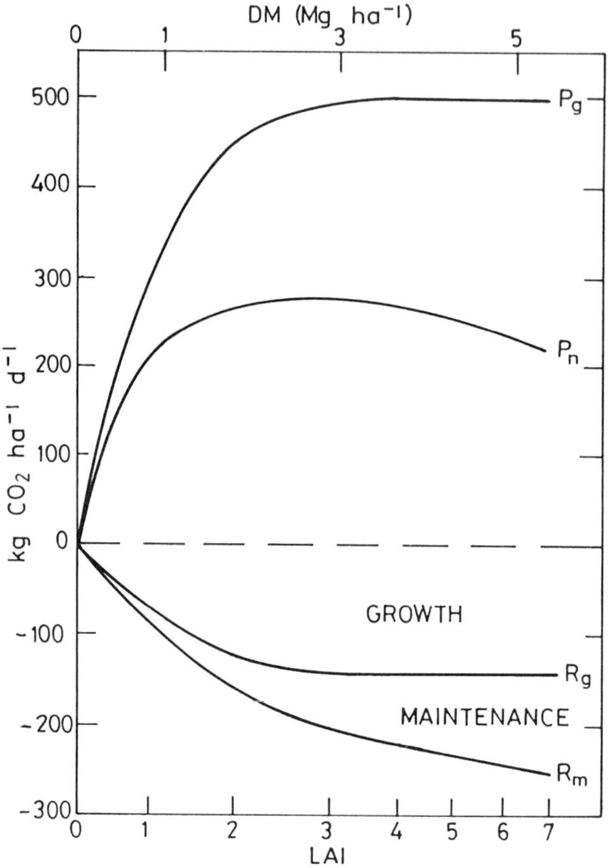

Fig. 2.5. Increase of gross photosynthesis (P_g), respiration (for growth and maintenance) (R), and net photosynthesis (P_n) with leaf area index (LAI) of field-grown white clover. (From Spitters, 1986).

a complex of anatomical, physiological and biochemical characteristics involved in the pathway of CO_2 fixation — Hatch–Slack photosynthetic pathway — which include most grasses (also cereal grasses) of tropical and subtropical areas. The light response curves of C_3 species have a lower initial slope and, possibly, a lower photochemical efficiency than those of C_4 species. These curves also show light saturation at 30–50% of full sunlight. On the other hand, in the absence of environmental or physiological limitations, light response curves for leaves of C_4 species do not saturate at full sunlight. The C_4 cycle appears to be more efficient in the conversion of radiant energy to DM, at least under high radiation and temperature, than the C_3 cycle (Ludlow et al., 1974; Monteith, 1978).

Field measurements of P_n, as affected by temperature and irradiance, for paspalum (C_4 species), perennial ryegrass and white clover (C_3 species) are shown in Figure 2.6. The rate of photosynthesis of ryegrass sharply decreased and that of paspalum increased with the increase of temperature, whereas that of white clover was least sensitive to temperature across all levels of irradiance. It should be mentioned, however, that

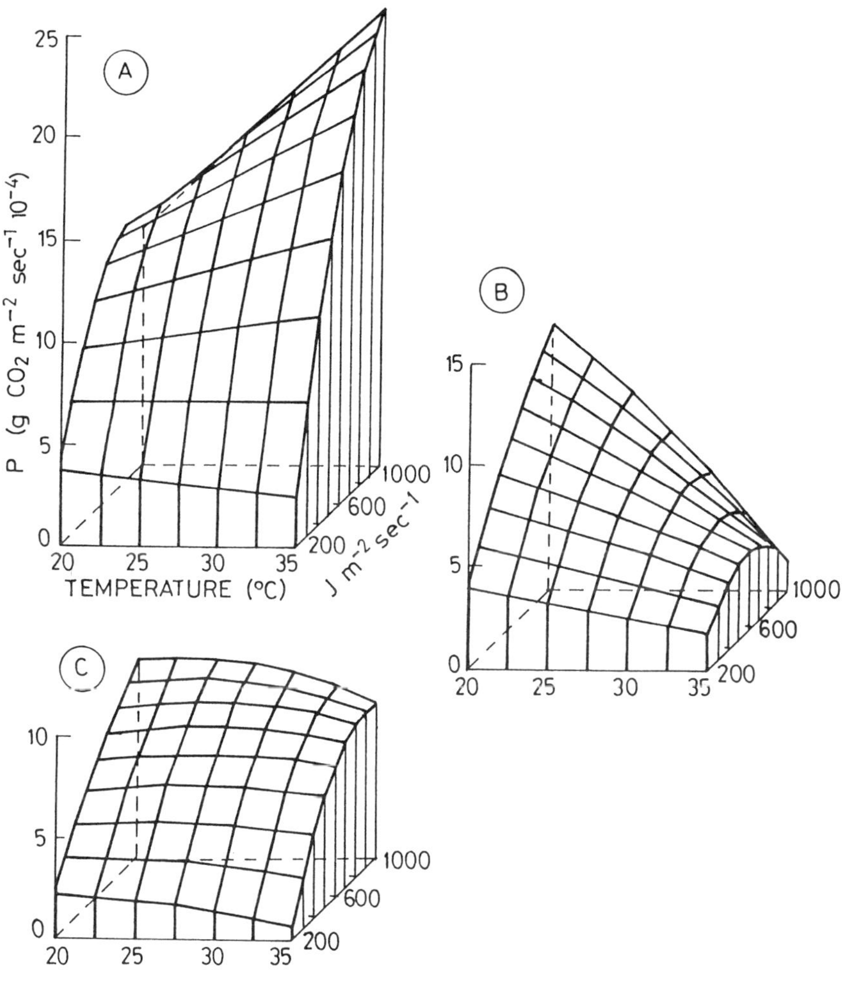

Fig. 2.6. Rates of net photosynthesis (P_n) for paspalum (A), perennial ryegrass (B) and white clover (C) as affected by ambient temperature and global irradiance. The data were collected in the field at Kyabram (Victoria, Australia) from well-watered swards, so there were few data points at either high temperature/low radiation or at low temperature/high radiation. (Adapted from Mason et al., 1987).

growth responses to temperature are quite genotype dependent, specifically of white clover.

2.2.3 Canopy Development and Growth

The stages through which the plant community proceeds during its life cycle reflect the phases through which the individual plant proceeds during its life cycle. However, inter-plant competition for space, i.e., light, can affect plant morphology. The effect of plant density on canopy structure is discussed for different forage crops in Part Two.

The DM accumulation of a forage corn crop is presented as an example in Figure 2.7A. This accumulation follows the typical trend of a sigmoidal curve. After seedlings emerge, growth proceeds exponentially until the canopy intercepts most of the light falling on it, i.e. until the foliage forms a 'closed' green surface. After canopy closure DM accumulation continues in a linear fashion. Finally, little or no further increase in DM takes place as a result of processes associated with maturation and leaf senescence.

Daily growth per unit ground area, called *crop growth rate* (CGR), follows a curve which rises, plateaus and falls, respectively (Fig. 2.7B). These phases of canopy development are usually referred to as the *seedling* or *juvenile phase*, the *closed-canopy phase*, and the *maturation phase*, respectively. Each of these developmental phases of corn is of different duration. The major contribution to DM accumulation occurs during the closed-canopy phase, i.e., when most of the incoming light is intercepted by the standing crop. Thus, total DM production is determined largely by the magnitude of the growth rate during this closed-canopy phase and its duration. The specific growth patterns of each of the major forage crops are discussed in Part Two.

Juvenile or Seedling Phase

Above-ground growth is characterized by the formation of leaves and by an increase of leaf area and corresponding LAI. As long as individual plants are small and develop without mutual interference, daily DM accumulation increases by a constant proportion tending toward the exponential. The duration of this period of exponential growth is dependent mainly on the number of established seedlings (or the number of tillers capable of regrowth after defoliation) per unit of ground area, and on growing conditions. High seeding rate and favorable growth conditions increase CGR but shorten the duration of the seedling phase, since the

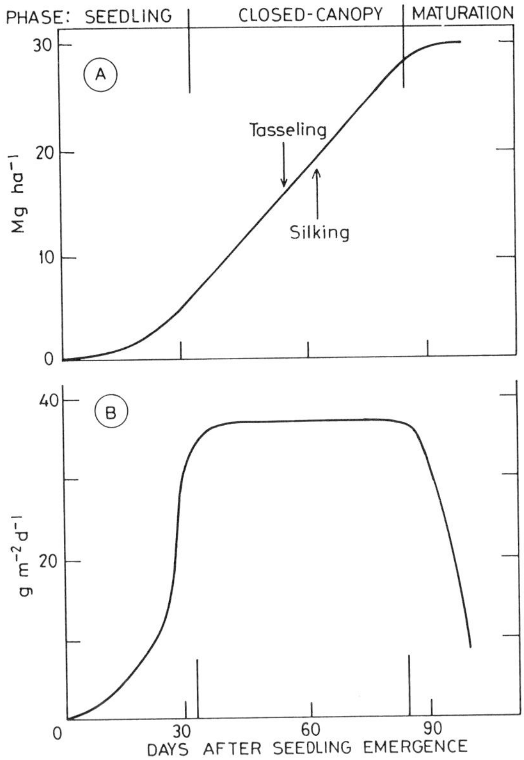

Fig. 2.7. Time course of dry weight accumulation (A), and crop growth (B) rate during development phases in trickle-fertigated forage corn. Bet Dagan, Israel. (Adapted from Yanuka et al., 1982).

ground is covered more rapidly with foliage compared with a low seeding rate or a small number of regrowing tillers and less favorable growing conditions. Higher than usual seeding rates are often recommended when high DM yields are required within a relatively short period of time or when crop rotation necessitates earlier harvesting.

Closed Canopy

This phase starts as soon as the foliage forms a closed green crop surface which results in mutual shading. Leaf and stem production proceeds in a linear fashion, i.e., CGR is constant after having reached a maximum value. Typical average growth rates during this phase are 20 g DM m^{-2} d^{-1} for C_3 plants (such as wheat, alfalfa, temperate grasses) and 30 g DM m^{-2} d^{-1} for C_4 crops (corn, sorghum and tropical grasses) under optimal growing conditions. The magnitude of the growth rate and the duration of growth are major determinants of the total DM yield. The end of the closed-canopy phase is attained as soon as DM accumulation slows down and CGR declines.

Maximum recorded short-term growth rates of some irrigated forages are shown in Table 2.1. A comparison of these rates demonstrates consistent differences between C_3 and C_4 crops. However, a number of factors, such as leaf age, canopy structure, respiratory load, temperature, irradiance, nutrient and moisture supply, may mask the differences in photosynthetic pathways. Significantly, it is of considerable importance to mention that C_4 crops have superior water use efficiency, which is discussed in Chapter 3.

TABLE 2.1

Maximum recorded short-term growth rates of some irrigated forage crops

Crop and CO_2 pathway	Location	Crop growth rate ($g\ m^{-2}\ d^{-1}$)	Reference
High-density corn, C_4	Davis, Calif.	51	Loomis and Williams (1969)
Forage corn, C_4	Bet Dagan, Israel	60	Yanuka et al. (1982)
Corn, C_4	San Camilo, S. Peru	50	Versteeg (1985)
High-density corn, C_4	Camden, NSW, Australia	59	Pearson et al. (1977)
Sudan grass, C_4	Davis, Calif.	51	Loomis and Williams (1969)
Rhodesgrass, C_4	Bet Dagan, Israel	30	Dayan et al. (1981)
Alfalfa, C_3	Davis, Calif.	23	Loomis and Williams (1969)
Annual ryegrass, C_3	Ramat Hanegev, Israel	20	Sagi (1992)

Maturation

This phase is characterized by a decrease in growth rate of the above-ground parts, which is associated mostly with reproductive development, and in cereal grasses with desiccation and the onset of rapid leaf senescence.

2.2.4 Mineral Nutrition

Plant nutrition management plays a vital role in the maintenance of high yield levels and forage quality. Its importance cannot be overestimated. Insufficient available plant nutrients during periods of sustained growth can affect yield and water use efficiency dramatically.

Fertilizer strategies to meet the nutrient requirements of forage crops require knowledge of:

(i) the availability of specific elements in the soil;

(ii) the amount and the rate at which these elements are taken up by the plants which become components of plant and animal products; and

(iii) the possible loss of elements through leaching below the root zone or volatilization.

The dynamics of these and associated processes are often very complex and are not discussed in detail within the context of this book. Reviews of mineral nutrition and soil fertility (and fertilizers) are provided by, e.g., Tisdale et al. (1985) and Marschner (1986); and those of alfalfa, corn and C_4 grasses by Lanyon and Griffith (1988), Arnon (1975) and Allan Jones (1985), respectively. Interpretation manuals of plant analysis of Australian crops and pastures are provided by Reuter and Robinson (1986) and Benton Jones et al. (1991). Aspects of mineral nutrition and fertilizer use of specific crops are discussed in Part Two.

Nutrient Availability in Soil

The nutrient-supplying capabilities of soil and the ensuing nutrient concentration in the soil solution vary with time and with soil type. For nutrients to be available, the amount of readily soluble nutrients in the soil profile and the rate of nutrient replenishment have to be known. Schofield (1955) suggested that the nutrient concentration in the soil solution can be characterized by an intensity factor, and the fraction adsorbed to the solid phase or bound to labile organic compounds can be characterized by a quantity factor as summarized in Figure 2.8 (after Marschner, 1986). The rate of replenishment from the labile pool to the soil solution around the root has to be extremely high for P and K. That of nitrate nitrogen ($N\text{-}NO_3$) depends on the nitrification of ammonium (NH_4) or on the mineralization of organically bound N. As the rate of N replenishment is often too low to cover the demand of fast-growing plants, the supply must be maintained either by root exploration of the subsoil or by split applications of N fertilizer. The intensity by which N is supplied from soil organic matter is much more difficult to assess than the quantity of bulk minerals such as P and K and microelements. Some approximations have been used in a

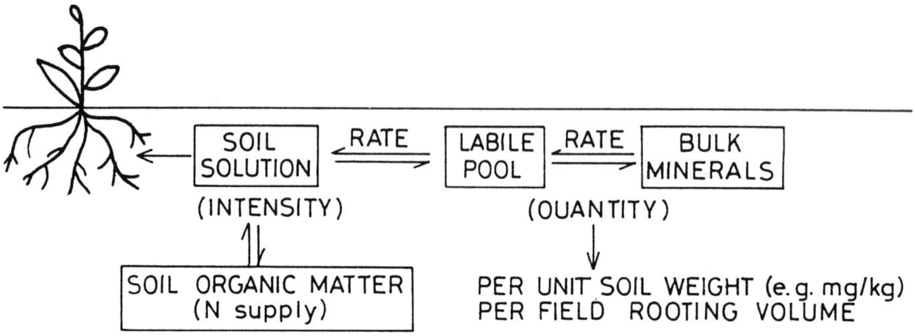

Fig. 2.8. Concept of the intensity–quantity ratio of soil nutrient availability. (From Marschner, 1986).

fertilization model for corn by Hadas et al. (1979).

Years of soil fertility research, related laboratory and crop response studies resulted in the development of various plant and soil testing procedures. These procedures provide a quite dependable prediction of nutrient needs of crops for growth or for maintenance of a high production level. In most areas where irrigation is practised, soil-test laboratories recommend fertilizer practices for different soils and crops. The procedures used in formulating recommendations by different laboratories are not necessarily identical. Using established standard values of nutrient levels in leaves, a comparison is made between the results of laboratory analyses with one or more known values or ranges in order to assess the plant's nutritional status. Another system of plant analysis interpretation is called DRIS (Diagnosis and Recommendation Integrated System), a method using ratios of element contents to establish a series of values that will identify the most to the least deficient elements (Benton Jones et al., 1991).

Nutrient Uptake by Crops

Removal of nutrients from the soil is the result of uptake and translocation by plants during growth. Usually concentrations of elements within the plant are more directly related to herbage growth than are concentrations in the soil (Pearson and Ison, 1987). Concepts of 'nutrient sufficiency', 'maintenance' or 'critical' levels for given elements in crop plants are employed in attempts to establish the composition level below which growth stress may begin and above which the yield will not increase further even with additional uptake (luxury consumption). Such discriminating parameters can be useful diagnostic tools for evaluating fertilizer practice or for estimating total nutrient uptake. Caution is required, however, because interpretations of plant analysis data mostly revolve around the question of how accurate is the 'ideal' or 'normal' composition of a given plant species when its growth or yield is at an optimum. Such information is derived mainly from the results of fertilizer rate experiments, but may be insufficient if forage quality also has to be considered.

To predict the optimum rate of fertilizer application, an estimate of the removal of nutrients by the crop is needed. If the approximate forage yield and the target (i.e. sufficiency) concentration of elements in the yield are known, then yield of elements can be reasonably estimated by multiplication. Target concentration implies the concentration of elements to be maintained during the period of maximum growth, which corresponds with the closed-canopy phase.

Nutrient Loss through Volatilization and Leaching

Volatilization is the result of the denitrification process, i.e., the biological reduction of NO_3-N to molecular N, which subsequently escapes to the atmosphere. Although the exact nature of the denitrification process is but poorly understood, it occurs primarily under low oxygen conditions such as in moist soils with high water-holding capacity and inside aggregates which are water-saturated (Terman, 1979). In the reverse process, nitrification, NH_4 is oxidized to NH_3 by soil microorganisms. One of the main factors influencing the direction and magnitude of these reduction oxidation processes is the soil moisture content (Kafkafi, 1973).

The distribution and percolation of fertilizer elements in the profiles of irrigated soils are associated with the behavior of the ions as related to conditions of fixation and adsorption, and are functions of the individual ions and of the nature of the environment. Generally there is little movement of P and K except under specific exchange conditions and on shallow soils. Well-planned irrigation scheduling must ensure that the water reaches the root zone. Under such conditions the hazards of NO_3 loss by leaching and pollution are minimized.

2.3 CONCLUSIONS

Yield formation is the product of many complex physiological processes affected by genetical, morphological and environmental factors. Dry matter accumulation in leaves and stems (sometimes in roots), during vegetative growth are the principal yield fractions harvested from most forages. In other forage crops, emphasis is placed on the content of highly digestible kernels derived from reproductive growth as part of the whole crop yield. The diversity in plant morphology and phasing of phenological development in time and space among crops and cultivars facilitates the selection and mode of utilization of irrigated forages.

Although water accumulation is the major process contributing to growth, photosynthesis is the basic rate-determining process in DM production, provided that the crop is supplied with adequate water and mineral nutrients. Warm season grasses (including forage corn and sorghum) with the C_4-type photosynthetic pathway, attain higher growth rates and use water more efficiently in warm climates than temperate grasses and legumes.

Maximum light interception by leaf canopies should be maintained as long as practical. Sustained herbage production requires the maintenance of vigorous regrowth after defoliation which depends on the presence of viable buds below the level of defoliation. Damage to stem or crown buds often results from wheel traffic of harvesting equipment, over-irrigation or poaching by grazing animals.

Fertilizer strategies to meet the nutrient requirements of forage crops require information on the availability of specific elements in the soil, and the amount and rate at which these elements are taken up. Vegetative growth commands higher elemental concentrations in the soil than reproductive growth. The maintenance of adequate nutrient levels in the soil solution at any time during the growing period avoids potentially detrimental fluctuations in the mineral supply to vigorously growing plants. Possible loss of elements through leaching below the root zone or volatilization should be prevented.

REFERENCES

Allan Jones, C. 1985. C_4 Grasses and Cereals: Growth, Development and Stress Response. John Wiley & Sons, New York, NY. 419 pp.

Arnon, I. 1975. Mineral Nutrition of Maize. Int. Potash Inst., Bern. 452 pp.

Benton Jones, J., Mills, H.A. and Wolf, B. 1991. Plant Analysis Handbook. Micro-Macro Publ. Inc., Athens, GA. 213 pp.

Davies, A. 1988. The regrowth of grass swards. pp. 85–127. In: M.B. Jones and A. Lazenby (Eds.) The Grass Crop. Chapman and Hall, London, UK.

Davies, I. 1973. Regrowth characteristics of an S.23 perennial ryegrass sward defoliated at early stages of development. J. Agric. Sci. Camb. 80: 1–10.

Dayan, E., Keulen, H. van and Dovrat, A. 1981. Tiller dynamics and growth of Rhodes grass after defoliation: A model named TILDYN. Agro-Ecosystems 7: 101–112.

Dovrat, A., Dayan, E. and Keulen, H. van. 1980. Regrowth potential of shoots and roots of Rhodes grass (*Chloris gayana* Kunth) after defoliation. Neth. J. Agric. Sci. 28: 185–199.

Eastin, J.D., Haskins, F.A., Sullivan, C.Y. and Van Bavel, C.H.M. (Eds.) 1969. Physiological Aspects of Crop Yield. Am. Soc. Agron. Madison, WI. pp. 396.

Evans, L.T. (Ed.) 1975. Crop Physiology. Cambridge Univ. Press, Cambridge, UK. 374 pp.
Gates, D.M. 1962. Energy Exchange in the Biosphere. Harper and Row Biol. Monogr., New York, NY. 151 pp.
Hadas, A., Bar-Yosef, B., Kafkafi, U. and Yaniv, Y. 1979. A model to determine N.P.K. fertilization recommendation under field conditions. Pamph. Agric. Res. Org., Bet Dagan 217. 57 pp.
Hay, R.K.M. and Walker, A.J. 1989. An Introduction to the Physiology of Crop Yield. Longman Sci. Tech., London. 292 pp.
Hyder, D.N. 1972. Defoliation in relation to vegetative growth. pp. 304–317. In: V.B. Younger and C.M. McKell (Eds.) The Biology and Utilization of Grasses. Academic Press, New York, NY.
Kafkafi, U. 1973. Nutrient supply to irrigated crops. pp. 177–188. In: B. Yaron, E. Danfors and Y. Vaadia (Eds.). Arid Zone Irrigation. Ecol. Stud. 5, Springer Verlag. Berlin.
Keulen, H. van and Wolf, J. (Eds.) 1986. Modelling of Agricultural Production: Weather, Soils and Crops. Pudoc, Wageningen, the Netherlands. 479 pp.
Kipnis, T. 1977. Morphological and physiological aspects of regrowth of Rhodes grass (*Chloris gayana* Kunth) after cutting. Ph.D. thesis, Hebrew University of Jerusalem, Rehovot, Israel. 127 pp.
Lanyon, L.E. and Griffith, W.K. 1988. Nutrition and fertilizer use. In: A.A. Hanson, D.K. Barnes and R.R. Hill (Eds.). Alfalfa and Alfalfa Improvement. ASA, CSSA and SSSA Monogr. 29: 334–372.
Loomis, R.S. and Williams, W.A. 1969. Productivity and the morphology of crop stands: Patterns with leaves. pp. 27–47. In: J.A. Eastin, F.A. Haskins, C.Y. Sullivan and C.H.M. Van Bavel (Eds.) Physiological Aspects of Crop Yield. Am. Soc. Agron., Madison, WI.
Loomis, R.S., Williams, W.A., Duncan, W.G., Dovrat, A. and Nunez, F. 1968. Quantitative descriptions of foliage display and light absorption in field communities of corn plants. Crop Sci. 8: 352–356.
Ludlow, M.M. 1978. Light relations of pasture plants. pp. 35–49. In: J.R. Wilson (Ed.) Plant Relations in Pastures. CSIRO, Melbourne, Australia.
Ludlow, M.M., Wilson, G.L. and Heslehurst, M.R. 1974. Studies on the productivity of tropical pasture plants. V. Effect of shading on growth, photosynthesis and reproduction in two grasses and two legumes. Aust. J. Agric. Res. 25: 425–433.
Marschner, H. 1986. Mineral Nutrition of Higher Plants. Academic Press, New York, NY, 674 pp.
Mason, W., Kelly, K., Blaikie, S. and Stockdale, R. 1987. New directions for irrigated pastures. Proc. 14th Austr. Agron. Conf. (Melbourne, Australia). pp. 100–117.
Metcalfe, D.S. and Nelson, C.J. 1985. The botany of grasses and legumes. pp. 52–63. In: M.E. Heath, R.F. Barns and D.S. Metcalfe (Eds.) Forages, 4th ed. Iowa State Univ. Press, Ames, IA.
Milthorpe, F.L. and Moorby, J. 1974. An Introduction to Crop Physiology. Cambridge Univ. Press, Cambridge, UK. 202 pp.
Monteith, J.L. 1969. Light interception and radiative exchange in crop stands. pp. 89–111. In: J.D. Eastin, F.A. Haskins, C.Y. Sullivan and C.H.M. Van Bavel (Eds.).

Physiological Aspects of Crop Yield. Am. Soc. Agron., Madison, WI.
Monteith, J.J. 1978. Reassessment of maximum growth rates for C_3 and C_4 crops. Exp. Agric. 14: 1–5.
Pearson, C.J., Dawbin, K.W., Muldoon, D.K. and Campbell, L.C. 1977. Growth and quality of tropical forages in a temperate climate. Aust. J. Exp. Agr. Anim. Husb. 17: 991–994.
Pearson, C.J. and Ison, R.L. 1987. Agronomy and Grassland Systems. Cambridge Univ. Press, Cambridge, UK. 169 pp.
Reuter, D.J. and Robinson, J.B. (Eds.) 1986. Plant Analysis: An Interpretation Manual. Inkata Press, Melbourne, Australia. 218 pp.
Robson, M.J., Ryle, G.J.A. and Woledge, J. 1988. The grass plant — its form and function. pp. 25–83. In: M.B. Jones and A. Lazenby (Eds.) The Grass Crop. Chapman and Hall, London, UK.
Sagi, M. 1992. MSc. thesis (unpublished). Hebrew University of Jerusalem, Israel.
Schofield, R.K. 1955. Can a precise meaning be given to available phosphorus? Soils Fert. 28: 373–375.
Spitters, C.J.T. 1986. Principles of Crop Production. Dept. Theor. Prod. Ecol. Agric. Univ. Wageningen, the Netherlands, 114 pp.
Taiz, L. and Zeiger, E. 1991. Plant Physiology. Binjamin/Cummings Publ. Co. Redwood City, CA. 558 pp.
Terman, G.L. 1979. Volatilization losses of nitrogen as ammonia from surface-applied fertilizers, organic amendments, and crop residues. In: N.C. Brady. (Ed.) Adv. Agron. 31: 189–223.
Tisdale, S.L., Nelson, L. and Beaton, J.D. 1985. Soil Fertility and Fertilizers. MacMillan Publ. Co., New York, NY. 754 pp.
Trenbath, B.R. and Angus, J.F. 1975. Leaf inclination and crop production. Field Crops Abstr. 28: 231–244.
Versteeg, M.N. 1985. Factors influencing the productivity of irrigated crops in southern Peru in relation to prediction by simulation models. Ph.D. thesis, Agric. Univ. Wageningen, The Netherlands, 181 pp.
Warren Wilson, J. 1967. Maximum yield potential. pp. 34–56. In: Transition from Extensive to Intensive Agriculture with Fertilizers. Proc. 7th Coll. Int. Potash Inst. (Tel Aviv, Israel), pp. 34–56.
Warndorff, M., Dovrat, A. and Kipnis, T. 1985. Quantifying tiller re-growth of hybrid pennisetum canopies. Neth. J. Agric. Sci. 33: 367–376.
Wit, C.T. de. 1965. Photosynthesis of leaf canopies. Agric. Res. Rep. 663. Pudoc, Wageningen, the Netherlands. 57 pp.
Woodward, F.I. and Sheehy, J.E. 1983. Principles and Measurements in Environmental Biology. Butterworths, London.
Yanuka, M., Leshem, Y. and Dovrat, A. 1982. Forage corn response to several trickle irrigation and fertilization regimes. Agron. J. 74: 736–740.

Chapter 3

WATER RELATIONS AND IRRIGATION

Y. Cohen

3.1 INTRODUCTION

The total water required by a given crop through either direct evaporation from the soil or by transpiration during the growing season, is generally similar in regions that have similar climatic conditions. Forage yield in comparable climatic conditions is generally higher if the water supply can be controlled by irrigation, than in un-irrigated forage. In arid, semi-arid, and sub-humid regions that have periods of drought, irrigation is a prerequisite for economic yield and stability of forage production throughout the season. In arid regions annual precipitation is generally <200 mm, and irrigation is the primary source of water for crops. However, forage crops are generally less common in such regions because of the high cost of water. Forage crops can be grown in semi-arid regions receiving 400–500 mm of precipitation annually, but without irrigation low and inconsistent yields are common. In sub-humid regions irrigation is generally not necessary for optimal crop production, but a dry period may decrease the yield if water cannot be supplemented.

In a semi-arid climate, irrigation must provide most of the crop water requirements during the dry period to meet water losses by the crop. Semi-arid regions are characterized by a high rate of water loss because of high solar radiation, large vapor pressure deficits, and frequent heat advection from the surrounding non-irrigated area. Some tropical climates have wet and dry seasons with favorable temperatures for crop growth throughout the year, providing that irrigation is applied during the dry season. A mediterranean-type climate with generally wet and cool winters enables cultivation of certain forage crops with little or no irrigation, but the summer is warm and rainless. In this type of climate, storage of water

applied water. The excess of the applied water is accounted for by other terms in the water balance equation (percolation, change in soil water content, runoff).

Four typical functions relating DM to applied water may be seen in Figure 3.2. The functional relationship between DM and seasonal irrigation water quantity are estimated from measurements of water quantity and the associated yield (Bresler, 1992). The best fitted results of DM forms alternative production functions of different types, e.g., sigmoid type, parabolic type, exponential (Mitscherlich) type, and piece-wise linear type. The figure shows the scattered sampling points and the best estimated lines.

If irrigation is reduced, DM production will be limited because T will not reach its maximum (Stewart et al., 1983). On the other hand increasing the amount of applied water will not affect the T rate but it will produce more surface runoff or deep percolation and, as drainage increases, yield will no longer be correlated with applied water (Jensen et al., 1980). IRR-WUE would be consistent for a given field and crop if the amount of water supply is below the maximum T. But when the control of water application is not sufficient and an excess water is given, the IRR-WUE may be dependent on soil types, field layouts, irrigation system, proportion of covered ground, and wetted soil volume (Sammis, 1981; Guitjens, 1990). The relationship of crop production to irrigation also depends on the salinity of the soil and irrigation water, the uniformity of the irrigation applications, the spatial variability of the soil physical properties, specific crop variety characteristics and crop cultural practices (e.g., weed and pest control, soil fertility, plant population, row spacing, and planting date). Seasonal IRR-WUE are commonly less than seasonal ET-WUE's because of different amount of water losses.

Engineering water use efficiency (EN-WUE) is the ratio of the amount of water stored in the crop root zone to the amount delivered for irrigation (Bos and Nugteren, 1978). If a new irrigation system is to be evaluated, this definition will be the most relevant.

3.2.2 Factors Affecting Water Use Efficiency

Due to a scarcity of water in arid and semi-arid regions and the cost of energy to supply the water to the crop, improving the T-WUE by the crop has been the goal of many research studies. WUE may differ between species. Diurnal changes in T-WUE occur in response to radiative flux and vapor pressure deficit, seasonal changes may be related to different temperature or growth stages. Plant species with the C_4 pathway of photosynthesis generally have lower CO_2 concentrations in the leaf interior and hence a larger CO_2 gradient between the interior leaf and the atmosphere, leading to a larger driving force for CO_2, than C_3 species. T-WUE is therefore generally higher in C_4 species. Fischer and Turner (1978) summarized many reports of the similarities between the transpiration ratios (TRs) (kg water transpired per kg DM produced) of plants with the same C metabolism pathway. They reported TRs of 667 and 303 kg of water per kg of DM for C_3 and C_4 plant species, respectively. Similar figures were reported by Stanhill (1986) for a large number of species. T-WUE may be changed by modifying climatic conditions, for example increasing air relative humidity would decrease the driving force for water vapor transport which may result in lower transpiration and higher WUE. However, for a crop in a moderate to high radiation environment, higher air humidity will result in a higher leaf temperature. Water vapor concentration in the leaf interior will rise as a consequence of the higher leaf temperature, partially offsetting the reduction in the driving force for water transport achieved by the increase in air humidity.

IRR-WUE can be improved by management; it may increase due to increasing the T fraction of the ET as a result of cultural practices such as weed control, timing of harvest, irrigation frequency, and minimizing soil evaporation and runoff. Favorable growing temperatures and lower ET in the spring resulted in a higher IRR-WUE than in the summer. Procedures for optimizing irrigation and water use efficiency must be determined experimentally for different environments because of the effect of climate and crop management on ET. It is widely accepted that in most crops, growth can proceed unimpaired and crop yield can be maximized only when the soil moisture potential remains high (and water remains readily available) continuously throughout the growing season. For most crops, keeping plant water potential high (by maintaining a moist irrigation regime) results in maximum production per unit area. The question remains as to whether it also results in maximum production per unit of water used. The answer appears to be that stressed plants do not use water more efficiently than well-watered plants (Rawlins and Raats, 1975). From the standpoint of DM production, either per unit of water used or per unit of land occupied, there seems to be no advantage in subjecting forage plants to water stress.

3.2.3 Water Use Efficiency in Forage Crops

ET-WUE of alfalfa varied according to climatic conditions, soil water regime (in which the efficiency was measured) season, location, annual ET,

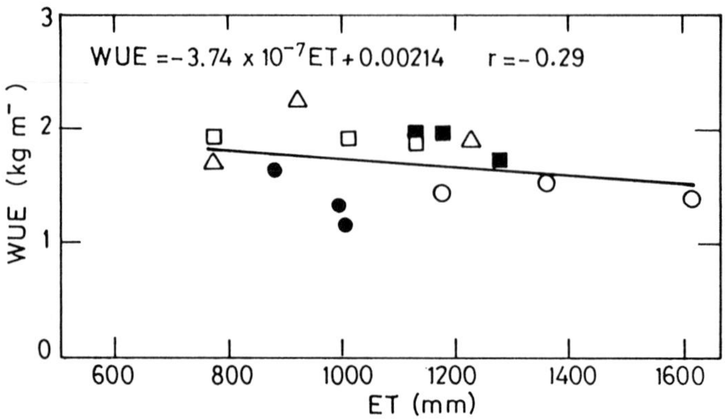

Fig. 3.3. Annual water use efficiency of alfalfa as a function of annual evapotranspiration. (After Guitjens et al., 1984).

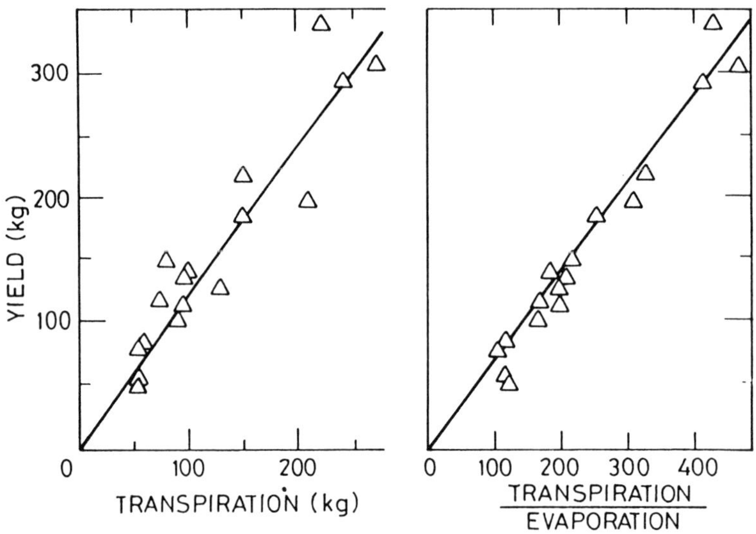

Fig. 3.4. Yield as a function of transpiration and the ratio of transpiration and evaporation. (After Hanks, 1983).

and growth cycle; WUE ranged from 0.5 to 3 kg m^{-3} (Metochis and Orphanos, 1981; Sammis, 1981; Guitjens, 1982; Donovan and Meek, 1983). Guitjens (1982) and Guitjens et al. (1984) found a negative linear relationship between annual WUE and water consumption in alfalfa (Fig. 3.3). The average annual WUE was 1.73 kg m^{-3}, with dry weight of hay rated at 12% moisture content for simulating field weight. Earlier, Peterson (1972)

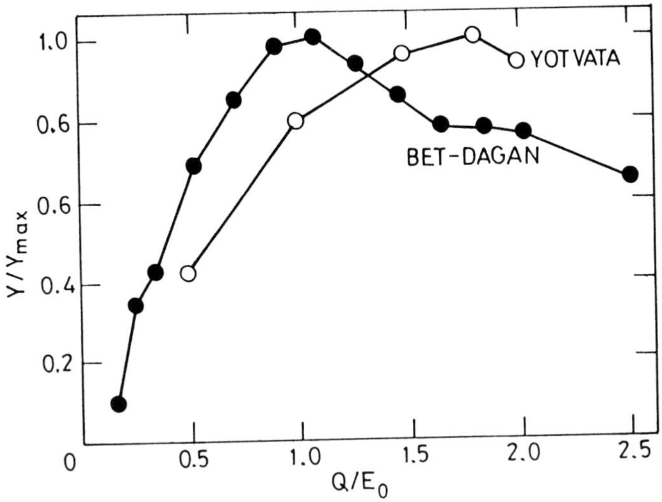

Fig. 3.5 Normalized corn yield (Y/Y_{max}) as a function of the relative amount of irrigation water (Q/E_0) at two regions in Israel characterized by different atmospheric evaporative demand. (After Cohen, 1992).

found a WUE of 1.47 kg m^{-3}. Hanks (1983) reviewed previous research on WUE with alfalfa grown in containers. There was a linear relationship between DM and T (a slope of 1.17 kg m^{-3}) but a better correlation was found between yield and the ratio of T to free water evaporation. The inclusion of free water evaporation removes the influence of differences among years (Fig. 3.4). Wenda and Hanks (1981) observed a linear relationship between both corn DM and ET and grain yield and ET. The seasonal ET-WUE, for well watered corn ranged from 1.2 kg m^{-3} DM in Bushland, Texas (Musick and Dusek, 1980) to as high as 2.2 kg m^{-3} in the southern Negev region of Israel (Hillel and Guron, 1973). However, the slope of the ET–yield relationship was dependent on environmental conditions and crop growth stage. Conditions of high atmospheric evaporative demand may decrease T-WUE as well as IRR-WUE because of high vapor pressure deficit, as explained above, and because of high direct evaporation. Figure 3.5 presents a normalized corn yield as a function of the relative amount of irrigation water (Q/E_0, where Q is applied water amount and E_0 is accumulated pan evaporation value) at two regions in Israel: Bet-Dagan (relatively moderate climatic conditions) and Yotvata in the Arava valley (extremely high evaporative demands). The accumulated pan evaporation for the entire growing period was 375 and 496 mm at Bet-Dagan and Yotvata, respectively. Yield increased steadily with increasing Q/E_0, but maximum yield at Yotvata was obtained at a ratio of

Q/E_0 nearly 80% higher than that at Bet-Dagan (Cohen, 1992).

A critical problem with any crop–water production function is the extent to which it can be generalized for different sites. Sammis (1981) developed water production functions for alfalfa in an effort to assess the transferability of production functions. He found that data from different sites generally fall on the same line, with the variation being as high from year to year as from location to location, and concluded that the same function may be employed at different sites. On the other hand, from an experiment with several irrigation treatments applied to three alfalfa varieties during two years, Hanks and Retta (1980) reported a linear relationship between ET and yield for both years; however, the effect of the location did differ between years. Adoption of any water input–DM production function (IRR-WUE) from one site to another may therefore be misleading, since daily and seasonal water use and plant growth are influenced by climate and cultural practices.

For maximum DM production, expressed either per unit of water or per unit of land, it is generally desirable to avoid water stress by maintaining a uniformly high water potential in the soil root zone. There is no evidence that alfalfa or other forage crops have growth stages which are particularly sensitive to water stress (Downey, 1972). Musick and Dusek (1980) observed 67% and 94% reductions in ET-WUE of corn when water deficits were imposed. Similar data were obtained by Eck (1986). The literature on irrigation emphasizes the importance of maintaining high soil water status at the time of anthesis, to avoid a reduction in yield.

3.3 ESTIMATING CROP WATER USE

3.3.1 Objectives of Irrigation Scheduling

Proper irrigation scheduling can reduce irrigation costs and increase crop production. To obtain maximum yields it is usually necessary to schedule irrigations in order to maintain a non-stressed soil-water status for much of the growing season. At the same time over-irrigation must be prevented in order to obviate water losses, leaching of salts and fertilizers below the root system, and reduction of the root zone aeration below optimal levels. In general, irrigation intervals are determined by the following parameters: crop species, stage of growth, climatic conditions, and the water-holding capacity of the soil. Irrigation is scheduled to allow a slight depletion of the soil water content and mild stress, but avoid yield reduction. The strategy of minimizing the number of irrigations per season, which results in larger irrigation amounts per irrigation, may better match the available labor situation and convenient farm water management. However, extended irrigation intervals may result in a significant decrease of T rate due water stress in the soil, especially when soil water-holding capacity is low. Irrigation frequency in forage crops varies from 4 to 14-d intervals. Over-irrigation will lessen irrigation efficiency and yield response to other management practices such as weed control. Deficient irrigation, on the other hand, will result in crop water stress and less yield response to other management factors, even though irrigation efficiency remains high because of less runoff and deep percolation. Technical and farming aspects of irrigation scheduling were comprehensively discussed earlier by Heermann et al. (1990).

3.3.2 Soil Water Balance

Irrigation requirements may be determined by many techniques with commercially available equipment of different accuracy and complexity. The technique should determine the timing and the amount of irrigation water to apply. Since soil water is the source for ET, and reduction of ET begins when the supply of water from the soil to the plants becomes limited, the most common method to estimate the irrigation requirement is according to the rate of change in soil water content. The method based on soil water balance which can be described in the following equation:

$$ET = SW + P + I - D + R \qquad [6]$$

where ET = evapotranspiration, SW = change in soil water content, P = precipitation, I = irrigation, D = drainage and R = runoff. Recording precipitation and irrigation should be routine procedures. Quantification of the drainage and runoff is difficult in the field, but can be determined by hydrological methods. To obtain data on changes in soil water content, periodic measurements are done by the gravimetric method or neutron probes technique.

The gravimetric method for determining SW is labor-intensive but needs only little investment in equipment. The neutron meter measurements can give the rate of change in soil moisture content between dry and wet conditions, and provide information for determining optimal and deficient soil water ranges. The use of the neutron probes method is more convenient and accurate than the gravimetric method, since the measurements are repeated at the same location in the soil. Calibration of the neutron probes appears to be the major source of error. Spatial variability of the soil in a heterogeneous field is another serious drawback to obtaining reliable information. A detailed description of soil water measurement is given by Campbell and Mulla (1990).

The measurements of soil water content provide information on readily available soil moisture: when this drops, ET and DM production decrease. Irrigation should take place before a reduction in DM production occurs as a result of decreasing ET. The amount of water application depends on the amount of depleted soil water in the root zone, soil water capacity, and irrigation efficiency. The soil water-holding capacity has to be determined at least once.

3.3.3 Soil Water Status

The most common method for monitoring soil water potential for irrigation is the tensiometer measurement. The measurement is simple, and recently new instruments were developed in which the potential in the water columns is expressed in electrical units which can be automatically stored in a computer. The water in the tensiometer cavitates at tensions around 85 KPa. This limits the measurement of soil water potentials to a range between 0 and –85 KPa. However, even though the range of the tensiometer is only a small fraction of the total range of water potentials over which plants can extract water from soil, in most forage crops irrigation is scheduled to eliminate drop in soil water potential to a level of tensiometer limit. The accuracy of the tensiometer measurement is also limited by the soil spatial variability and root system distribution which create sampling problems.

Another popular method for measurement of soil water potential is the use of electrical resistance sensors. These are usually made from gypsum, though fiberglass and nylon have also been used. The resistance of the sensor is related to the soil potential, however, a major problem with this method is that the ions in the soil solution can also affect the resistance. Detailed description of the above and other methods for measuring soil water potential were provided by Campbell and Mulla (1990).

3.3.4 Plant Water Status

Various parameters are used as indicators of plant water status; the most common is tissue or organ water potential. This parameter, which is described as the most commonly measured, is quantitatively linked to other parameters more closely tied to plant functions, and its reduction relative to well-watered controls can be correlated with productivity and yield (Hiler et al., 1972). Irrigation scheduling can be based on direct measurement of small stems or leaves using a pressure chamber. The method is convenient for field use and very rapid. The leaf is excised with a sharp razor blade, immediately covered and sealed in the chamber with only the minimum section of petiole protruding out. Pressure is applied until sap just appears at the cut end of most of the xylems as indicated by light reflection or glitter. A major source of error can be if the sample was not protected after excision and during the measurement. The time of day for these measurements is particularly important, since the lowest water potentials occurs at noon when the potential ET rate is highest.

Measurement of canopy temperatures was intensively used as a method for determining irrigation need in forage and other crops. This method is based on the assumption that as water availability to the plant becomes limiting, T decreases causing less evaporative cooling of leaves which results in increasing plant temperatures in relation to air temperature. Measurement of canopy temperature with an infrared thermometer has been used to evaluate stress conditions (Idso et al., 1981; Jackson et al., 1981; Reginato, 1983). This method involves a crop water stress index (CWSI), described by:

$$CWSI = 1 - (ET/ETp) \qquad [7]$$

where ET_p is the potential evapotranspiration. Abdul-Jabbar et al. (1985) showed a linear decline in alfalfa yield with a change in CWSI from zero to one.

3.3.5 Meteorological Approach

Estimating ET directly from the measurement of atmospheric demand for water, circumvents the necessity of repeatedly quantifying all components in the soil water balance equation. Micrometeorological methods used for estimating atmospheric demand were comprehensively described by Hatfield (1990). Penman's equation (1948, 1963) has generally been accepted as a scientifically sound formulation for estimating ET. The equation is expressed as a combined function of radiation, air temperature, vapor pressure, and wind speed. Comparisons with lysimeter data have shown this equation to be both accurate and reliable. One limitation to the reliability of the equation occurs under conditions of advection. Hanks et al. (1971) found deviations of over 45% when warm dry air moved over an irrigated plot. The amount and frequency of meteorological data input for measurement of ET depend on the required accuracy. In most real-time irrigation scheduling applications, meteorological data are collected from automatic weather stations for a general regions or area. However, some studies have also used long-term average climatic data for scheduling irrigation. The limitation of using meteorological methods associates with measuring instruments and data collection. Accurate measurements are needed of vapor pressure and wind speed. These limitations have prompted the development of empirical approximation for irrigation scheduling rather than the generalized theoretical approach.

3.3.6 Empirical Equations

Empirical equations that consider only a single parameter such as solar radiation or temperature have been used. The Priestley–Taylor (1972) and Jensen and Haise (1963) methods estimate ET as a function of solar radiation and temperature. These two methods have been most successful in humid areas where the convective energy is either relatively small or nearly constant.

Another empirical method which integrates the effect of radiation, temperature and relative humidity on evaporation is that employing Class A pan. This method measures the amount of water which is evaporated from a pan with known dimensions. The ease of use, simplicity, and low cost have prompted the evaporation pan method to be adopted widely. Pruitt (1966) described in detail the adjustments required in pan evaporation data to account for the surrounding area. Doorenbos and Pruitt (1977) used this method in developing their guidelines for determining crop water use. Thom et al. (1981) made a comprehensive analysis between the Penman equation and pan evaporation and concluded that the latter was adequately described by a combination equation with an adjustment in the psychometric constant and wind function for the pan.

Pan evaporation data have been used to develop curves for potential consumptive use of water by crops. Relationships between meteorological parameters and pan evaporation have been developed for locations around the world. Several techniques have also been introduced into the evaporation pan to enable automation of the readings (Phene and Campbell, 1975; Asrar et al., 1982; McKinion and Trent, 1985). Limitations on the use of evaporation pans are related to the environment in which they are located. The relatively large volume of Class A pans acts as a heat sink which causes a delay in its response to changes in climatic parameters. Pan evaporation data may be acceptable when they are based on a measurement period of 5–10 days.

Crop evapotranspiration (ET_c) correlates with atmospheric evaporative demand. It is useful to relate ET_c to a standard index of ET of grass which has the potential of maximum transpiration rate and known as a reference crop (ET_0). ET_0 is defined as the amount of water transpired per unit time by a short, closed grass canopy of uniform height, and never short of water. ET_c may also be related to pan evaporation (E_p). While E_p has been used as an index for evaporative demand for many years, ET_0 is considered a more realistic parameter for use in irrigation scheduling. Many studies used the ET_0 to develop specific crop factors, hereafter referred to as crop coefficients (K_c), that relate ET_c to ET_0. The ET_0 is usually determined from meteorological data collected at weather stations which are located in well defined environments. Climatic factors and the canopy structure of the crop determine the rate of ET as long as the surface of the field is effectively fully wet. Under water stress conditions, i.e., when the soil surface begins to dry, and when the stomata begin to close, ET drops to below its potential rate. During the daytime when there is adequate light for photosynthesis, stomata of healthy and actively growing crops, not short of water, are fully or nearly fully open for most of the day time, and T is at the potential rate. As soil water is depleted and the water potential of the crop drops to low values, stomata close as a protective mechanism, and transpiration is reduced. Stomata also tend to close when growth conditions are unfavorable, when the crop matures and the leaves senesce. Due to the effect of stomatal resistance on crop ET, it is usually lower than ET_0.

Calculation of ET_0 is a procedure commonly used in irrigation management programs and is related to one of the empirical methods generally the evaporation pan method. Doorenbos and Pruitt (1977) and Jensen (1973) discussed variation among species and its effect on irrigation management in assessing crop water requirements. They showed that K_c is a good criterion for assessment of crop water requirements.

3.4 LOW WATER QUALITY

3.4.1 Resources

The scarcity of water resources in many regions has increased the use of high salinity water from aquifers, brackish, industrial, domestic and sewage effluent resources. The widespread use of fertilizers adds a further salt burden to the soil. Health regulations limit the use of the above types of water for horticultural crops, but there is less restriction to their use for forage crops if no animal health hazard is expected. In some arid and semi-arid regions the use of low quality water has lent major and increasing importance to the role of irrigation in ameliorating the soil's chemical composition. The use of treated municipal sewage effluent for irrigation of forage crops when other water resources are limited is an attractive and practical reuse option. In addition to receiving supplementary water from this source, the crop can utilize some of the components of the waste-water such as organic matter, N and P. These water resources, when used for irrigation, are no longer discharged to surface or underground water bodies; therefore, the pollution of natural water resources is considerably reduced.

The accumulation of heavy metals in crops receiving waste-water has been of concern largely because of the high metal content of industrial waste-waters or sludge, although the metal content of crops irrigated with waste-water has not been found to be elevated (Campbell et al., 1983). High NO_3 levels in ground water have been associated with waste-water irrigation projects where the application of N exceeds the crop's removal of this element. On the other hand, irrigation with waste-water was found to increase the quality of forage crops.

Any assessment of the suitability of a saline water for irrigation must be made while keeping the following in mind: how much salinity can the crop tolerate in the soil water; how much leaching will be achieved; and what level of salinity will result in the soil water from the use of this irrigation water. The suitability of irrigation water should be evaluated on the basis of the specific conditions of use, including the crops grown, soil properties, irrigation management, cultural practices, and climatic factors. The prevailing criteria of irrigation water quality and their associated potential hazards to crop growth are:

(i) Salinity — the general salt effects on crop growth thought to be largely osmotic in nature and related to total salt concentration. These effects are generally evidenced by retarded growth, producing smaller plants with fewer and smaller leaves;

(ii) Sodicity — the effect of an excessive amount of exchangeable sodium in the soil on soil permeability, soil structure deterioration, and a direct toxic effect of exchangeable sodium on plants specifically sensitive to sodium (Rhoades and Loveday, 1990).

In many of the scattered small areas of irrigation, the natural rate of ground water diffusion beyond the root zone appears sufficient to prevent the buildup of high water tables and consequent salinization. However, when poor-quality ground water has been used, salinization problems have occurred. Many forage species are resistant to salinity, remove significant amounts of salt by uptake, and can be irrigated with contaminated water without posing a health hazard (Bole et al., 1985). For these reasons forage crops play a major hydrological role in eliminating low-quality water infiltration to aquifers.

3.4.2 Application

The use of low-water quality for irrigation has also necessitated the development and application of advanced irrigation systems and techniques for proper and economic water application. These systems enable efficient water use while also utilizing marginal waters. An example of one such advanced irrigation system is the sub-surface trickle irrigation that can economize greatly on water while enabling economic yields under conditions of high water and soil salinity (Phene, 1988; Bar-Yosef et al., 1989; Oron et al., 1990;). Effluent quality must be taken into account when selecting the irrigation method to be used. Surface irrigation methods can utilize low-quality effluent, whereas sprinkler and particularly drip irrigation require higher-quality effluent, because of the danger of suspended solids clogging orifices of the irrigation equipment. This problem can be overcome by installing filters at the head of the irrigation system. When surface irrigation is used, the water supply canals are constructed of soil of such permeability that considerable loss due to seepage occurs. Such seepage is often a major cause of the development of high water tables and excessive soil salinity in irrigated lands. Seepage losses can be reduced by compacting the canal floor and walls or by lining them with less permeable materials.

Salinity level in the root zone varies in time and depth, depending on the degree to which water is depleted between irrigations and on the leaching fraction (Rhoades and Merrill, 1976). Crop yields have been shown to be closely correlated with the level of salinity in the root zone and the growth stage at which the root system was exposed to this level (Bresler, 1987). When the soil is completely wet, water is used from the upper, less saline root zone, until the total water stress (matric and osmotic potential) becomes greater in the upper than in the lower root zone, at which time water is used from the lower root zone. Once this situation is reached, salinity effects on crop growth become increasingly apparent, unless irrigation intervenes. Since salinity effects are predominantly water stress effects, there are obvious implications for irrigation management. For

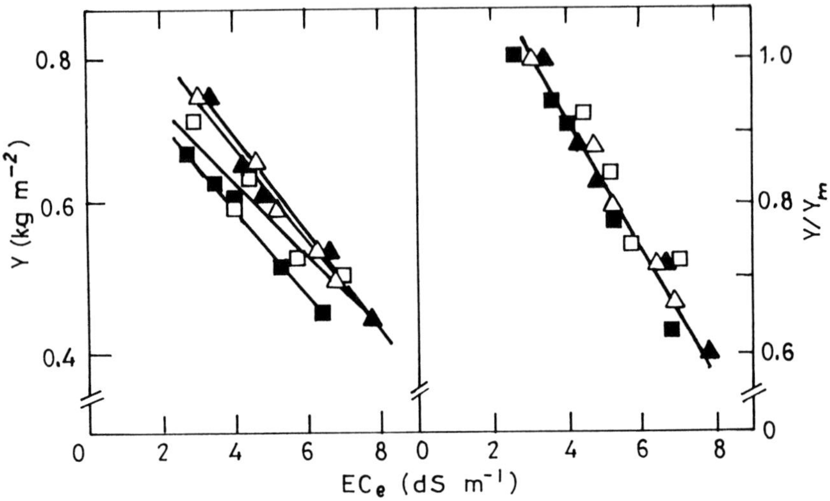

Fig. 3.6. The relation of absolute (Y) and relative (Y/Y_m) yield of corn stover dry matter as a function of soil salinity (EC_e) at four irrigation intervals: 3.5 d (▲), 7 d (△), 14 d (□), 21 d (■). (After Shalhevet et al., 1986).

example, plants can tolerate higher levels of salinity under conditions of low matric stress resulting from high-frequency irrigation.

In general, improvements in salinity control come from improvements made in on-farm irrigation efficiency, i.e., by providing the appropriate amount of water at the appropriate time with high uniformity of application. The ideal irrigation scheme would provide water more or less continuously to the plant to match ET losses and to maintain the water content in the root zone within narrow limits commensurate with adequate aeration and minimum loss in deep percolation for leaching. By this means the salinity of the soil water is prevented from increasing significantly between irrigation events as ET proceeds. The availability of the water to the crop is maximized since the matric and osmotic potentials are maximized. To achieve such an ideal system requires delivery of the water to the field on demand at appropriate flow rates and volumes.

The best means of controlling soil and water salinity is by efficient irrigation with adequate but minimum leaching; and drainage management that maintains the net flux of water in the soil downward, and minimizes off-site pollution. When highly saline water is used, the relation between crop yield and seasonal amount of applied water is dependent on the design of irrigation systems and the planning of irrigation management. This relationship is complicated by salinity, because sufficient water must be applied to provide for leaching in addition to ET.

The common recommendation is to increase the irrigation frequency when saline water is used (Ayers and Westcot, 1976). While there are few concrete data to support this recommendation, there is also some clear information to the contrary. An increased irrigation frequency may result in an upward shift of the peak of salt concentration distribution with depth. Furthermore, as irrigation becomes more frequent, the evaporation component of ET increases, leading to additional water application and increase in the salt load (Shalhevet, 1992). The net result of these two processes is difficult to predict; experiments with several crops (including forage crops) have shown no advantage of increasing frequency of irrigation with saline water (Fig. 3.6). Some results, actually showed a damaging rather than beneficial effect of increasing frequency of irrigation with saline water (Wagenet et al., 1980).

3.4.3 Leaching

In attempting to minimize the water supply to crops, soil salinity may increase due to insufficient leaching from a continued partial wetting of the root zone; this may be particularly critical when irrigating with low-quality water (Shmueli, 1973). Furthermore, under an extreme limit of water supply, crop yield may be reduced to a level at which the financial return will not be sufficient to cover the expenses of the total input.

Plant water uptake and evaporation create the potential for an upward flow of water into the root zone. The root zone and the soil surface may become salinized by the process of water movement from deeper soil layers. This effect may be considerable, especially where shallow saline water tables are present. The removal of salt from the root zone to maintain the soil solution at a salinity level compatible with the cropping system is referred to as maintaining salt balance (Rhoades and Loveday, 1990). The fraction of infiltrated water that passes through the root zone is called the leaching fraction (L) i.e., D_{dw}/D_{iw}, where D_{dw} and D_{iw} are the depth of drainage water and irrigation water, respectively. The L can be estimated for steady-state conditions from EC_{iw}/EC_{dw}, where EC_{iw} and EC_{dw} are the electrical conductivities of the irrigation water and the drainage water, respectively. A salt balance relation in the root zone can be expressed by the following equation:

$$D_{dw}/D_{iw} = EC_{iw}/EC_{dw} \qquad [8]$$

In this equation the equivalent water depth (D) is substituted for volume. It expresses the approximate equality, under the given conditions, between the leaching fraction ($L = D_{dw}/D_{iw}$) and the ratio of the irrigation to drainage water salinity (EC_{iw}/EC_{dw}). Thus, by varying the fraction of applied water that is percolated through the root zone, it is possible to control the concentration of salts in the drainage water within certain limits, and hence to control either the average or the maximum salinity of the soil water in the profile at some desired level. The relationship between average root zone salinity of the soil-saturation extract and leaching fraction for two irrigation water concentration is shown in Figure 3.7.

The leaching requirement is the fraction of the total applied water that must percolate through the root zone. This fraction is a function of salt concentration in the irrigation water. For example, with water containing 1,000 mg l^{-1} or less of dissolved solids, the leaching requirement for most crops is < 0.05 (van Schilfgaarde et al. 1974; Maas and Hoffman, 1977). Thus, even with the most efficient irrigation management, if uniform area distribution is obtained the leaching requirements are likely to be minimized. Under severe water limitations, production of salt-tolerant forage species may not be reduced by irrigation with water of high salt concentration if there is proper management of leaching. As indicated by Rhoades (1977), irrigation with water containing 6,300 mg l^{-1} of salt at L = 0.25 should enable high yield of relatively salt-sensitive crops. Specific soils and

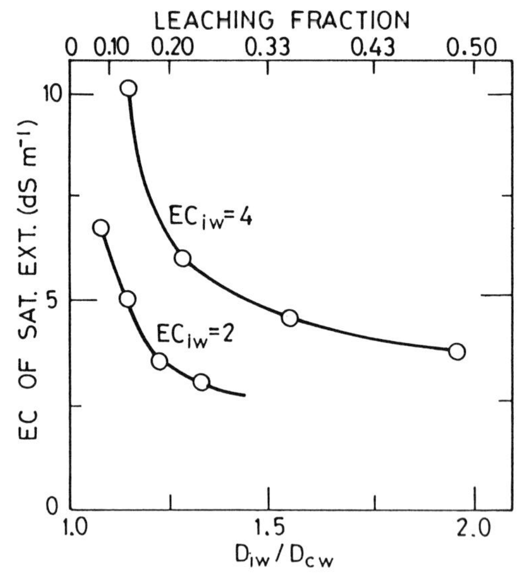

Fig. 3.7. Relation between average root zone salinity, expressed as electrical conductivity (EC) of the soil-saturation extract, and leaching fraction of two irrigation water concentrations. (After Bower et al., 1969).

crops add constraints to leaching efforts: for example, clay soil with limited rate of water infiltration, or alfalfa which develops root rot and scalding if the crowns are covered with ponded water. It may be concluded that L depends primarily on the crop's sensitivity to salinity, the salt concentration in the irrigation water and in the soil, physical characteristics of the soil, climatic conditions, and water extraction distribution (Bresler et al., 1982).

The traditional approach to the control of root-zone salinity (the steady-state approach) has been to add a constant leaching fraction to the irrigation, depending on water quality and water extraction distribution. The more useful and advanced approach — "delay leaching and apply it when needed" — has several advantages (Meiri, 1990):

(i) rainfall in many regions causes all or at least some part, of the leaching and slows down soil salinization;

(ii) leaching synchronized with crop salt tolerance can reduce salinity damage during sensitive stages and allow salt accumulation during tolerant stages; and

(iii) more concentrated drainage water and more effective leaching, which also reduce the hazard of a high water-table buildup, cause secondary soil salinization.

Typically, the leaching fraction exceeds the leaching requirement in most irrigation projects, providing an opportunity to improve irrigation management so as to reduce leaching fraction without harming crop production. Commensurate with a reduction in the leaching fraction, the volume of applied irrigation water is reduced and hence the added salt which must be leached is also reduced. Minimizing the leaching fraction results in the smallest possible total salt load in the drainage flow from the root zone because it maximizes the precipitation of salts and minimizes mineral weathering and dissolution of previously deposited salts.

3.4.4 Yield Response

The relationship between soil salinity as expressed by the EC of the soil saturation extract and relative yield is shown in Figure 3.8. The general mathematical relationship may be expressed by the equation:

$$Yr = 100 - (EC_e - t)\ S$$

where t is the threshold salinity in which the yield just begins to decline and S is the rate of yield decline with increasing salinity. Yr is the relative

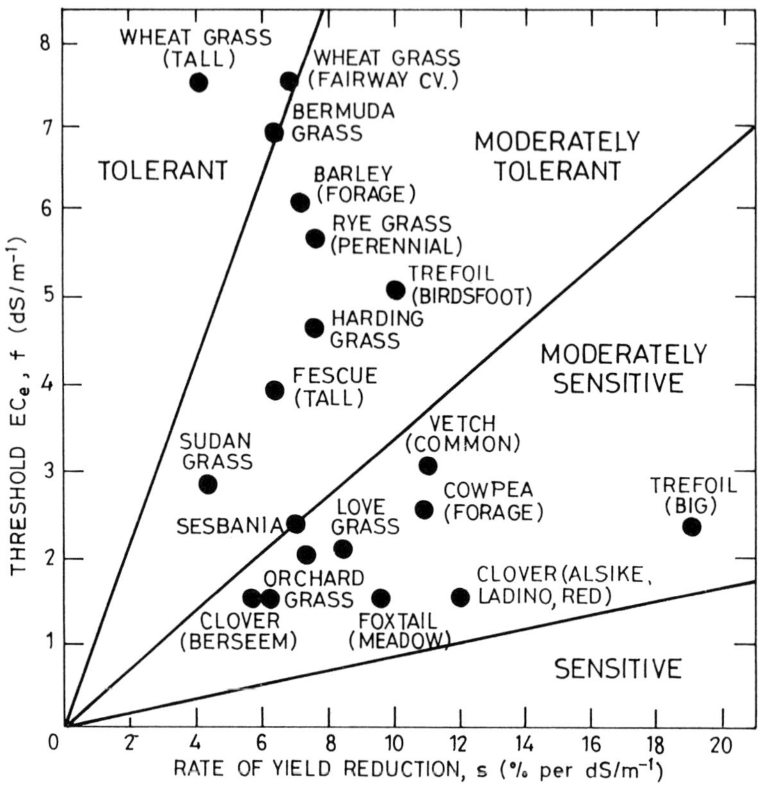

Fig. 3.8. Salt tolerance ratings of forage crops, for explanation of s and t, see text. (After Maas, 1986).

yield in percent and EC_e is the mean soil salinity expressed as the conductivity of the soil saturation extract. Salt tolerance is characterized by the values of both the plateau and slope.

Water stress resulting from saline soils can reduce the net CO_2 assimilation and yield of the crop in different ways and to different extents, depending on its timing relative to the growth stage and its severity. The leaf expansion stage is very sensitive to even mild water stress. When water stress occurs during the canopy development phase, canopy size may be reduced to a small fraction of that of the well-watered control. The impact on assimilation and productivity would be dramatic, since only a small fraction of the sunlight would be captured by the canopy for photosynthesis. Another effect of stress on leaves is the accelerated senescence of older leaves. Under field conditions, senescence appears to be slightly more sensitive to water stress mostly at the maturation stage, the impact on yield can be considerable. The number of flowers per unit of land area is the second most sensitive to water stress, this is most likely the result of the extreme sensitivity of leaf and canopy growth to water stress. Geometrically, the number of sites for the formation of reproductive structures on a plant is clearly related to the size of its canopy. Physiologically, the number of fruiting sites must be dictated to some degree by the assimilates potentially available.

Excess salinity in the root zone adversely affects the growth of established plants by a general reduction in growth rate. Salt stress essentially increases the energy that must be expended by the plant to extract water from the soil and to make the biochemical adjustments necessary to grow, relative to the non-saline condition. Besides the above-mentioned effects on established crops, salinity also adversely influences crop establishment, especially under conditions of furrow irrigation. In fact, obtaining a good stand of plants is often the factor that most limits crop production in saline areas. Once the stand is established, management risks are generally substantially reduced. The problem of reduced seedling establishment is due in part to the generally lower salt tolerance of seedlings compared with established plants. Additionally, the problem is enhanced because the seeds or small seedlings are exposed to locally excessive salinity in the seed-bed that often occurs there as a result of evaporation.

The use of low-quality water, requires information on a crop–water production function under these specific conditions to develop optimum irrigation management strategies. The crop–water production function will be unique for each crop, as crops vary in their sensitivity to salinity. Models developed for these conditions (Letey et al., 1985) assumed a linear relationship between yield and ET. Strong experimental evidence supports a linear relationship between yield and ET for forage crops or for total top weight of other crops. Measured commercial yields and above-ground DM production compared well with yields computed on the assumption that relative crop yield is equivalent to relative T. Both measured and computed results indicated that irrigation water quality and quantity, rather than irrigation frequency, influenced DM production of grass (Bresler and Hoffman, 1986).

The effects of soil salinity and soil water content on crop yield are interacting variables and thus it is difficult to separate crop response to these two soil variables. Since salts are not transferred to the atmosphere as water evaporates from the soil or transpires from the plant, they are concentrated in the soil solution (Bresler and Hoffman, 1986). These processes change both the osmotic and matric potentials of the soil solution, which have been shown to be additive in their effect on T (Childs and Hanks, 1975). However, in conducting experiments on water requirement, crop production functions, and salt tolerance, researchers have minimized the effect of matric potential when studying salinity effects, and made the osmotic potential negligible when plant response to soil water is of interest. In irrigated agriculture where salinity is a hazard, matric and osmotic potentials must be considered simultaneously. For example, if insufficient amounts of saline irrigation water are applied, matric and osmotic potential will both be low, and T and growth will be reduced. It is generally assumed that the major effect of soil salinity in reducing crop yield is via decreased plant water uptake.

The common goal of an agricultural enterprise is to maximize profit. This is particularly true under conditions of water scarcity or where drainage systems must be installed. For example, some areas do not have natural disposal sites for drainage water and thus it may be necessary to construct evaporation ponds. In some cases productive land must be sacrificed for the building of evaporation ponds. Under these conditions optimal water management could be significantly different from management to maximize crop production.

3.5 IRRIGATION SYSTEMS

3.5.1 Objectives

The primary objective of an irrigation system is to supply the plant with the correct amount of water at the proper time. The use of a given irrigation system calls for a specific approach to irrigation water management, since water influences most production factors — from tillage and planting, to seed germination, root extension, nutrient management and uptake, leaching of hazardous salt to maintain a favorable salt balance and, most important, plant growth and yield formation.

The key factor in the supply of water to the crop is management decisions regarding irrigation and uniformity of application. The grower may directly control the former and strongly influence the latter. Water–DM functions should economically justify water application to the crop. For maximum IRR-WUE, the irrigation system should perform with minimum water loss both by percolation and by direct evaporation from the soil surface. High efficiency requires both good water-management decisions (e.g. when to irrigate, how much water to apply) and irrigation systems designed and maintained to achieve high uniformity of applied water. Irrigation systems must minimize hazards to plant and soil, such as: soil salinization, burns on leaves and other plant tissues, soil erosion, destruction of soil structure, and creation of a crust which increases runoff by preventing water penetration into the soil. Generally, irrigation losses result from the deep percolation of water below the root zone and the runoff of excess water from the end of the field (sprinkler systems also involve spray evaporation and wind drift losses).

Deep percolation occurs because irrigation water cannot be applied with complete uniformity to a field, regardless of the method used. Some areas of the field receive more water than others, and when the infiltrated amount exceeds the water-holding capacity of the root zone, water travels beyond the depth where it can be extracted by the plant. The extent of deep percolation depends on the amount of water applied, the storage capacity of the root zone, and the hydraulic conductivity of the soil. Spatial variability of soil infiltration properties has only a minor effect on the extent of deep percolation. On the other hand, uniformity depends mostly on the irrigation system. Properly designed and maintained pressurized systems allow greater control of infiltrated amounts and uniformity than do surface systems, which have intake rates that vary over the field and limit distribution uniformity.

The design of field irrigation systems consists primarily of four factors: the total amount of required irrigation water, irrigation intervals, irrigation method, and the layout of the irrigation system. In this section surface and pressurized irrigation system properties will be dealt with in respect to their application in forage crops. The major types and comparisons of different irrigation systems in the USA have been described by Kruse et al. (1990).

3.5.2 Surface Irrigation

Application of water by means of furrows, flood, border or basins using a non-pressurized system is known as surface irrigation. The investment for equipment and for a system to transport water to the field is relatively low, but the surface irrigation method requires a high labor input. The system is characterized by a high percentage of water loss, problems with uniformity of water distribution, and with controlling water and fertilizer regimes in the root zone. For these reasons this method is almost unused in regions with limited water sources. On the other hand, the method is widely used in forage crops when water resources are less limited.

The uniformity of infiltrated water in the field depends on the rate at which water advances across the field and the spatial variability of soil intake rates. The opportunity time for infiltration is usually highest at the inflow end of the field, decreasing with distance down the run. However, infiltration opportunity time differences between the two ends of the field can be reduced by increasing the inflow rate at the top of the field, although higher inflow rates may lead to excessive runoff and over-irrigation.

A relatively new surface irrigation technique that has been shown to reduce differences in intake opportunity time is the application of water as a series of pulses rather than as a continuous flow. Intermittent application reduces the soil intake rate, which increases the rate of water advance. Studies of furrow-irrigated fields have shown that the water required for full advance can be reduced by 20–50% with pulse irrigation relative to continuous flow. Even using high non-erosive inflow rates for a rapid advance, deep percolation at the top of the field is likely under full-coverage irrigation (flood, border, or basin) when soil infiltration rate is high. In such cases single furrow would be an appropriate solution, because lateral movement of water in the soil would partly replace water infiltration to deep layers. Partial wetting of the soil surface by using the single furrow method will also reduce the extent of surface evaporation. Technical and agronomic aspects of surface irrigation are reviewed by Alvin Bishop et al. (1967) and Rawitz (1973).

3.5.3 Sprinkler Irrigation

The area under sprinkler irrigation has increased markedly in the last few years, as a result of mechanization, which minimizes labor. This system's main advantages are the very high degree of control of most of the important irrigation features, such as: water application rate; irrigation throughout the 24 hours of the day and night; time intervals between successive irrigations; allocation of a precise water amount per irrigation; and easy adaptation to automation. Specific management characteristics of sprinkler irrigation are as follows:

(i) It is easy to adjust sprinkler irrigation intensity and rate to the infiltration capacity of the soil during irrigation; this will prevent runoff and erosion of fertile soil.

(ii) Irrigation water quantities are flexible and can be controlled easily, by selecting sprinklers with appropriate discharge under optimal operative pressure and by proper spacing between sprinklers.

(iii) Has a high WUE.

(iv) May be used for frost protection and for humidifying the air in the crop's environment.

(v) Simple to operate and does not require any special expertise.

(vi) Saves labor but requires expensive investments; the inverse relationship between labor and investment depends on the type of system (i.e., stationary, portable, semiportable, solid set, mobile sprinklers, etc.).

(vii) Requires a relatively large amount of energy for optimal operation.

(viii) It is possible to irrigate on almost any topography without the need for leveling.

The uniformity of sprinkler-applied water depends on hydraulic design, system maintenance, and wind conditions during operation. Systems are usually designed so that the hydraulic pressure varies by no more than 20%; this results in discharge rates that should differ by less than 10%. In addition to variations in discharge rates, other factors that affect distribution are the spacing and height of risers, sprinkler and nozzle type, system pressure, and wind velocity.

The most common types of sprinkler systems can be specified according to their cost and field performance. Stationary (permanent) and solid-set systems cost more than other types initially, but once installed, the labor requirements are minimal; initial investment can be reduced at the expense of labor by alternate operation. In portable systems, the mains, sub-mains and laterals are all made of aluminum with gun (giant) sprinklers or conventional rotating sprinklers. Semi-portable, manually moved systems have buried sub-mains and light-weight portable laterals along which the sprinklers are evenly spread. Manual movement of laterals can be tedious and time consuming, particularly in heavy soils. For this reason, laterals mounted on wheels or skids are used, and towed across the sub-mains to their new settings. This may also be an alternative to supplement systems with giant sprinklers. Semi-portable sprinkler systems are the most widely used, they require fewer laterals and the initial investment is lower than with other systems. However, the labor required to remove the pipes from the field to facilitate farm operations may pose a problem.

Mobile irrigation methods are a common solution for precise water application to field crops. Unlike stationary high-intensity irrigation, where a circular ponded area is created on the soil surface, around the water source, a different geometry and consequently physical boundary conditions are developed when the water source is moved along the field. Where the water source is a single emitter, a ponded strip is created; this zone moves with the irrigator along the field. If, instead, there are multiple water sources along the irrigator wing, spread at distances from each other so as to create a continuous ponding zone, a plane ponded area is created and moves with the irrigator along the field. The single source will probably be employed where the seed bed is prepared as elevated rows, and below-canopy emitters when saline irrigation water is used and leaf burn is a possible hazard. The larger the discharge and the lower the hydraulic conductivity, the longer it takes the crop to approach its final size. An increase in the discharge rate expands the ponded area almost linearly which suggests that when the ponded area is large enough, the border effect on water absorption is small and the ponded area is affected mostly by soil hydraulic conductivity. Lower hydraulic conductivity causes a larger ponded area (Shani, 1992).

Mobile Sprinkler Machines

Labor scarcity on irrigated farms has accelerated the trend toward mechanized, mobile irrigation systems, such as center-pivot, traveling gun boom sprinklers, and traveling lateral systems.

Three types of mobile machines are used:

1. The *center-pivot system* consists of rotating sprinklers on a single lateral (usually 15 cm diameter) several hundred meters long, with one end anchored to a fixed pivot structure and the other end continuously moving around the pivot. Water is supplied from the sources to the lateral through the pivot. The lateral pipes with the sprinkler are supported on a drive unit with a power sensor mounted on it that drives the wheels on which the unit moves, at a predetermined speed.

2. *Traveling gun and boom sprinklers* have been changed from intermittently to continuously moving sprinklers mounted on trailers and fed by long flexible hoses. The major components of the traveling sprinkler system are pump, power unit, main or supply pipe, irrigation hose, traveler unit, and a high-capacity sprinkler mounted on a chassis to which a flexible hose is connected. The traveler unit is pulled along selected travel lanes by a cable wrapped on a rotating cable reel. The cable reel can be powered by a water turbine, water piston, or engine mounted on the traveler. The cable reel pulls the traveler through the field in a straight line, with the cable serving as an alignment guide. The traveling system is adaptable to many field sizes and shapes, with topography from level to rolling and irregular.

3. The *traveling lateral system* with sprinkler laterals that move continuously in a rectilinear fashion, was developed to irrigate corner areas, which are left unirrigated by a typical center-pivot system. The laterals are either fed from a flexible hose, or equipped with a traveling pumping plant to take water from a ditch. The traveling sprinkler laterals can irrigate a rectangular field completely, including corners. The flexible hose used for such systems is identical to that used for the traveling guns and boom sprinklers discussed above. A typical traveling sprinkler lateral would use a

single center mainline (or ditch). The lateral is situated to traverse half the field on one side of the mainline; when it reaches the end, its wheels are turned 90° so that the lateral can be returned to its original position. One estimate sets the labor requirement for these moves at approximately 6 man-hours per irrigation on a 65-ha field. Unlike the center pivot, the sprinkler spacing and discharge must be uniform throughout the entire length of the traveling lateral.

3.5.4 Drip Irrigation

The use of drippers as a means of irrigating large fields has become common practice in some locations of irrigated agriculture. However, the system has been established in forage crops to a limited extent only. Because of its high cost, this method is not applied in crops with relatively low income per unit area or water, including forage crops. Drip irrigation differs from other conventional irrigation methods in that it enables high-frequency irrigation, and thus has the potential to produce higher than usual yields. The long infiltration stage plays a dominant role in the irrigation cycle, and hence the soil's conductance properties are more important for drip irrigation design than the soil's capacity properties. Conversely, capacity is the main soil property considered for the design of sprinkler and flood irrigation systems, where a relatively long period of water uptake by plants through transpiration is the dominant stage in the irrigation cycle.

The main limitation to more extensive use of drip irrigation is its vast number of components, resulting in high costs for purchase and maintenance. The design problem is then to select the proper combination of dripper spacing and discharge for a given soil and crop. The particular advantage of the drip irrigation system lies in the fact that only the soil and roots are in contact with the water, the foliage remaining dry during the irrigation cycle. Thus the development of plant pathogens is minimized and damage to the leaves due to high salt content in the irrigation water is prevented. As a result the use of brackish water becomes possible for salt-sensitive crops.

Early studies demonstrated that plant roots tend to concentrate near the dripper. The introduction of plastic delivery systems and an improvement in the dripper structure enabled, for the first time, accurate delivery of water to individual plants over large areas, irrespective of the soil physical properties and chemical composition. The drip irrigation system has been reviewed by Dasberg and Bresler (1985). From virtual non-use in 1960, it was adopted world-wide and is still continuing to expand. When drip irrigation systems are properly designed, maintained and managed, they offer the greatest possible uniformity and the least irrigation losses. In practice, more than one-third of the drip irrigation systems have distribution uniformities of less than 70%. This has been ascribed primarily to filtration and water quality problems that cause emitter plugging. The potential achievable uniformities of drip systems approach 90%. Moreover, these systems offer irrigators the greatest control possible in the frequency and durations of irrigations. As such, they are ideally suited for use in restricted irrigation strategies that limit deficit periods to the least sensitive plant growth stages.

3.6 CONCLUSIONS

Forage yield is generally higher if the water supply can be controlled by irrigation, as compared with non-irrigated forage. Irrigation is a prerequisite for economic yield and stability of forage production throughout the season. In arid regions forage crops are generally less common because of the high water cost of water, although potential productivity under irrigation is high due to the high radiative flux. On the other hand, in sub-humid regions, due to the low vapor pressure deficit, crop production and water use efficiency, may be substantially higher than in arid regions. From the standpoint of dry matter production, either per unit of water used or per unit of land occupied, generally there is no advantage in subjecting forage plants to water stress.

To study the effects of irrigation on crop production in a given environment, a response curve of crop production to transpiration rate (transpiration-water use efficiency) must be studied. If data on the transpiration rate of the crops are not available, evapotranspiration figures may be collected more easily by a relatively simple means and can be used for studying the crop–water response curves. Generally, the adoption of any water input–crop production function from one site to another may be misleading, because daily and seasonal water use and plant growth are influenced by climate and by cultural practices.

Irrigation of forage crops is scheduled mostly by the changes in soil water content, but the use of soil and plant water status or empirical equations is also common. Forage crops are sensitive to deficient irrigation almost equally throughout the entire growth stage. In almost all forage crops irrigation is given mostly to allow a slight depletion of the soil water content and mild stress, but avoid yield reduction. In arid and semi-arid regions soils must also receive additional water beyond the water losses by the crops in order to leach accumulated salts from the root zone.

The use of low-quality water for irrigation of forage crops requires information on a crop-water function under these specific conditions to develop optimum irrigation management strategies. The crop-water production function will be unique for each crop, as crops vary in their sensitivity to salinity. The effects of soil salinity and soil water content on crop yield are interacting variables and thus it is difficult to separate the crop's response to these two soil variables.

The key factors in obtaining high efficiency of water applied to the crop are management decisions regarding irrigation and uniformity of application. Irrigation management may be controlled by the farmer on the basis of accurate knowledge of crop water requirements. Uniformity of water application can be considerably improved by introducing modern irrigation techniques.

REFERENCES

Abdul-Jabbar, A.S., Lugg, D.G. Sammis, T.W. and Gay, L.W. 1985. Relationship between crop water stress index and alfalfa yield and evapotranspiration. Trans. ASAE 28: 454–461.

Alvin Bishop, A., Jensen, M.E. and Hall, W.A. 1967. Surface irrigation systems. In: Irrigation of Agricultural Lands, R.M. Hagan, H.R. Haise and T. Edminster (Eds.) ASA, Monogr. 11: 865–884, Madison, WI.

Arkley, R.J. 1963. Relationship between plant growth and transpiration. Hilgardia 34: 559–584.

Asrar, J., Kunze, R.J. and Linuitt, D.E. 1982. Automating a class A evaporation pan for semi-continuous recording. Agric. Meteorol. 25: 275–281.

Ayers, R.S. and Westcot, D.W. 1976. Water quality for agriculture. F.A.O. Irrig. Drain. Pap. 29.

Bar-Yosef B., Sagiv, B. and Markovitch, T. 1989. Sweet corn response to surface and subsurface trickle fertigation. Agron. J. 81: 443–447.

Bole, J.B., Gould, W.D. and Carson, J.A. 1985. Yields of forages irrigated with waste water and the fate of added nitrogen-15-labeled fertilizer nitrogen. Agron. J. 77: 715–719.

Bos, M.G., and Nugteren, J. 1978. On irrigation efficiencies. Int. Inst. Land Reclam. Improve. Publ. 19 (2nd ed.). Wageningen, The Netherlands.

Bower, C.A., Ogata, G. and Tucker, J.M. 1969. Rootzone salt profiles and alfalfa growth as influenced by irrigation water salinity and leaching fraction. Agron. J. 61: 783–785.

Bresler, E. 1987. Application of a conceptual model to irrigation water requirement and salt tolerance of crops. Soil Sci. Soc. Am. J. 51: 788–793.

Bresler, E. 1992. Sprinkler irrigation management and optimization. Proc. of Binational China–Israel Workshop on Water Use Efficiency in Agriculture. J. Shalhevet, Liu Changming and Xu Yuexian (Eds.) (Beijing, China) pp. 224–251.

Bresler, E., and Hoffman, G.J. 1986. Irrigation management for soil salinity control: Theories and tests. Soil Sci. Soc. Am. J. 50: 1552–1560.
Bresler, E., McNeal, B.L. and Carter, D.L. 1982. Saline and Sodic Soils. Springer-Verlag, Berlin.
Campbell, W.F., Miller, R.W. Reynolds, J.H. and Schreeg, T.M. 1983. Alfalfa, sweetcorn, and wheat responses to long-term application of municipal waste-water to cropland. J. Environ. Qual. 12: 243–249.
Campbell, G.S., and Mulla, D.J. 1990. Measurement of soil water content and potential. In : Stewart, B.A. and D.R. Nielsen (Eds.) Irrigation of Agricultural Crops. Agron. Monogr. 30: 127–142. ASA, CSSSA, SSSA, Madison, WI.
Childs, S.W., and Hanks, R.J. 1975. Model of soil salinity effects on crop growth. Soil Sci. Soc. Am. Proc. 39: 617–622.
Cohen, Y. 1992. Factors determining irrigation demand in arid lands. Proc. Int. Symp. on Land Degradation and its Biological and Technological Rehabilitation in Drylands. Published by Arid Land Research Center (Tottori, Japan), pp. 71–79.
De Wit, C.T. 1958. Transpiration and crop yields. Versl. Landbouwkd. Onderz. 64.6.
Dasberg, S., and Bresler, E. 1985. Drip Irrigation Manual. Intern. Irr. Inform. Center, Bet Dagan. pp. 120.
Donovan, T.J., and Meek, B.D. 1983. Alfalfa responses to irrigation treatment and environment. Agron. J. 75: 461–464.
Doorenbos, J., and Kassam, A.H. 1979. Crop water requirements. FAO Irrig. Drain. Pap. 33.
Doorenbos, J., and Pruitt, W.O. 1977. Guidelines for predicting crop water requirements. FAO Irrig. Drain. Pap. 24 (Revised).
Downey, L.A. 1972. Water yield relations for non-forage crops. J. Irrig. Drain. Div., Am. Soc. Civ. Eng. 98: 107–115.
Eck, H.V. 1986. Effects of water deficits on yield, yield components and water use efficiency of irrigated corn. Agron. J. 78: 1035–1040.
Fischer, R.A., and Turner, N.C. 1978. Plant productivity in the arid and semiarid zones. Annu. Rev. Plant Physiol. 29: 277–317.
Guitjens, J.C. 1982. Models of alfalfa yield and evapotranspiration. J. Irrig. Drain. Div., Am. Soc. Civ. Eng. 108 (IR3): 212–222.
Guitjens, J.C. 1990. Alfalfa. In : B.A. Stewart, and D.R Nielsen, (Eds.) Irrigation of Agricultural Crops. Agron. Monogr. 30: 537–568, ASA, CSSSA, SSSA, Madison, WI.
Guitjens, J.C., Tsui, P.S. Connor, J.M. and Thran, D.F. 1984. Toward total water management. Proc. 12th Int. Congr. Irrig. Drain. Vol. 1(A), Q38: 169–184.
Hanks, R.J. 1974. Model for predicting plant yield as influenced by water use. Agron. J. 66: 660–665.
Hanks, R.J. 1983. Yield and water-use relationships: An overview. In: H.M. Taylor, W.R. Jordan and T.S. Sinclair (Eds.) Limitations to Efficient Water Use in Crop Production, ASA, CSSA and SSSA, Madison, WI.
Hanks, R.J., Allen, L.H. and Gardner, H.R. 1971. Advection and transpiration of wide-row sorghum in the central Great Plains. Agron. J. 63: 520–527.
Hanks, R.J., and Retta, A. 1980. Water use and yield relations for alfalfa. Bull. Utah Agric. Exp. Stn. 506: 1–8.
Hatfield, J.L. 1990. Methods of estimating evapotranspiration. In: B.A. Stewart and

D.R Nielsen (Eds.) Irrigation of Agricultural Crops. ASA, CSSA, SSSA Monogr. 30: 435–474.
Heermann, D.F., Martin,D.L. Jackson,R.D. and Stegman, E.C. 1990. Irrigation scheduling controls and techniques. In: B.A. Stewart, and D.R. Nielsen (Eds.) Irrigation of Agricultural Crops. ASA, CSSA, SSSA Monogr. 32: 509–535.
Hiler, E.A., van Bavel, C.H.M. Hossain, M.M. and Jordan, W.R. 1972. Sensitivity of southern peas to plant water deficit at three growth stages. Agron. J. 64: 60–64.
Hillel, D., and Guron, Y. 1973. Relation between evapotranspiration rate and maize yield. Water Resour. Res. 9: 743–748.
Idso, T.C., Jackson, R.D. Pinter, P.J. Jr., Reginato, R.J. and J.L. 1981. Normalizing the stress-degree-day parameter for environmental variability. Agric. Meteorol. 24: 45–55.
Jackson, R.D., S.B. Idso, R.J. Reginato, and P.J. Pinter. 1981. Canopy temperature as a crop water stress indicator. Water Resour. Res. 17: 1133–1138.
Jensen, M.E. 1973. Consumptive use of water and irrigation water requirements. American Society of Civil Engineers, NY.
Jensen, M.E., and Haise, H.R. 1963. Estimating evapotranspiration from solar radiation. J. Irrig. Drain. Div., Am. Soc. Civ. Eng. 89: 15–41.
Jensen, M.E., Harrison, D.S. Korven, H.C. and Robinson, F.E. 1980. The role of irrigation in food and fiber production. In M.E. Jensen (Ed.) Design and Operation of Farm Irrigation Systems. ASAE Monogr. 3. ASAE, St. Joseph, WI. pp. 15–41.
Kruse, E. G., Bucks, D.A. Von Bernuth, R.D. 1990. Comparison of irrigation systems in irrigation of agricultural crops. In: B.A. Stewart and D.R. Nielsen (Eds.) ASA, CSSA, SSSA Monogr. 30: 475–508.
Letey, J., Dinar, A. and Knapp. K.C. 1985. Crop-water production function model for saline irrigation waters. Soil Sci. Soc. Am. J. 49: 1005–1009.
Maas, E.V. 1986. Salt tolerance of plants. Appl. Agric. Res. 1: 12–26.
Maas, E.V., and Hoffman, G.J. 1977. Crop salt tolerance — current assessment. J. Irrig. Drain. Div., Am. Soc. Civ. Eng. 103(IR2): 115–134.
McKinion, J.M., and Trent, A. 1985. Automation of a Class A evaporation pan. Trans. ASAE 28: 169–171.
Meiri, A. (1990) Effective use of multi-quality water supplies for irrigation. Proc. 5th Int. Conf. Irrigation, Israel Agritech (Tel Aviv, Israel), pp. 165–185.
Metochis, C., and Orphanos, P.E. 1981. Alfalfa yield and water use when forced into dormancy by withholding water during the summer. Agron. J. 73: 1048–1050.
Musick, J.T., and Dusek, D.A. 1980. Irrigated corn yield response to water. Trans. ASAE 23: 92–98.
Oron, G., DeMalach, J. and Hoffman, Z. (1990) Sub-surface trickle irrigation of field crops with domestic effluent. Proc. 5th Int. Conf. Irrigation, Israel Agritech (Tel Aviv, Israel), pp. 243–257.
Penman, H.L. 1948. Natural evaporation from open water, bare soil, and grass. Proc. R. Soc. London 193: 120–146.
Penman, H.L. 1963. Vegetation and hydrology. Tech. Commun. Common. Bur. Soils, 53.
Peterson, J.B. 1972. Water relationships and irrigation. In: C.H. Hanson (Ed.) Alfalfa Science and Technology. Agron. Monogr. 15: 469–480. ASA, CSA, SSSA, Madison WI.

Phene, C.J. 1988. Underground drip system yields record tomato harvest. Irrig. J. 38: 12–14.

Phene, C.J., and Campbell, R.B. 1975. Automating pan evaporation measurements for irrigation control. Agric. Meteorol. 15: 181–191.

Priestley, C.H.B., and Taylor, U.J. 1972. On the assessment of surface heat flux and evaporation using large scale parameters. Mon. Weather Rev. 100: 81–92.

Pruitt, W.O. 1966. Empirical method of estimating evapotranspiration using primary evaporation pans. Proc. Conf. Evapotranspiration. ASAE, St. Joseph, MI. pp. 57–61.

Rawitz, E., 1973. Gravity irrigation. In: B. Yaron, E. Danfors and Y. Vaadia (Eds.) Arid Zone Irrigation. Spinger Verlag, Berlin. pp. 324–337.

Rawlins, S.L., and Raats, P.A.C. 1975. Prospects for high frequency irrigation. Science 188: 604–610.

Reginato, R.J. 1983. Field quantification of crop water stress. Trans. ASAE 26: 772–775.

Rhoades, J.D. 1977. Potential for using saline agricultural drainage waters for irrigation. Proc. Water Management, Irrigation and Drainage Symp. (Reno, NV), pp. 85–116.

Rhoades, J.D., and Loveday, J. 1990. Salinity in irrigated agriculture. In: Irrigation of Agricultural Crops. B.A. Stewart, and D.R. Nielsen (Eds.) Irrigation of Agricultural Crops. Agron. Monogr. 30: 1089–1142. ASA, CSSSA, SSSA, Madison, WI.

Rhoades, J.D., and Merrill, S.D. 1976. Assessing the suitability of water for irrigation: Theoretical and empirical approaches. FAO Soils Bull. 31: 69–109.

Ritchie, J.T. 1983. Efficient water use in crop production: Discussion on the generality of relations between biomass production and evapotranspiration. In: J.M. Taylor, W.R. Jordan and T.S. Sinclair (Eds.) Limitations to Efficient Water Use in Crop Production. ASA, CSSA, and SSSA, Madison, WI. pp. 29–44.

Sammis, T.W. 1981. Yield of alfalfa and cotton as influenced by irrigation. Agron. J. 73: 323–329.

Shalhevet, J., Vinten, A. and Meiri, A. 1986. Irrigation interval as a factor in sweet corn response to salinity. Agron. J. 78: 539–545.

Shalhevet, J. 1992. Using water of marginal quality for crop production: major issues. Proc. of binational China–Israel workshop on water use efficiency in agriculture. J. Shalhevet, Liu Changming and Xu Yuexian (Eds.) (Beijing, China) pp. 17–52.

Shani, U. 1992. The design of field irrigation systems as related to crop type and soil properties. Proc. Binational China–Israel workshop on water use efficiency in agriculture. J. Shalhevet, Lin Changming and Xu Yuexian (Eds.) (Beijing, China). pp. 195–210.

Shmueli, E. 1973. Efficient utilization of water in irrigation. In: B. Yaron, E. Danfors, and Y. Vaadia (Eds.) Arid Zone Irrigation, Ecological Studies 5, Springer-Verlag, Berlin. pp. 411–423.

Stanhill, G. 1986. Water use efficiency. Adv. Agron. 39: 53–85. N.C. Brady (Ed.) Academic Press, Inc.

Stewart, B.A., and Hagan, R.M. 1973. Functions to predict effects of crop water deficits. J. Irrig. Drain. Div., Am. Soc. Civ. Eng. 99: 421–439.

Stewart, B.A., Hagan, R.M. Pruitt, W.O., Hanks, R.J., Riley, J.P., Danielson, R.E. Franklin, W.T. and Jackson, E.G. 1977. Optimizing crop production through control

of water and salinity levels in the soil. Utah Water Res. Lab. Publ. PRWG. 151-1.
Stewart, B.A., Musick, J.J. and Dusek, D.A. 1983. Yield and water use efficiency of grain sorghum in a limited irrigation–dryland farming system. Agron. J. 75: 629–634.
Tanner, C.B., and Sinclair, T.R. 1983. Efficient water use in crop production: Research or re-search: In: H.M .Taylor, W.R. Jordan and T.S. Sinclair (Eds.) Limitations to Efficient Water Use in Crop Production. ASA, CSSA, and SSSA, Madison, WI. pp. 1–27.
Taylor, H.M., and Ashcroft, G.L. 1972. Physical Edaphology. Freeman and Co., San Francisco, CA.
Thom, A.S., Thony, J.L. and Vauclin, M. 1981. On the proper employment of evaporation pans and atmometers in estimating potential transpiration. Q. J. R. Meteorol. Soc. 107: 711–736.
Van Schilfgaarde, J., Bernstein, L. Rhoades, J.D. and Rawlins, S.L. 1974. Irrigation management for salt control. J. Irrig. Drain. Div. Am. Soc. Civ. Eng. 100 (IR3): 321–338.
Wagenet, R.J., Campbell, W.P., Bamatraff, A.M. and Turner, D.L. 1980. Salinity irrigation frequency and fertilization effects on barley growth. Agron. J. 72: 969–974.
Wenda, W.I., and Hanks, J. 1981. Corn yield and evapotranspiration under simulated drought. Irrig. Sci. 2: 193–204.

book, the term 'forage quality' is discussed and used here in its narrow sense, namely, *plant composition as related to potential nutritive value* — especially energy nutrients and their digestibility.

The evaluation of forage quality is obviously of great concern to both the grower producing forage for the market, and the dairy operator. The market price paid for conserved forages (e.g. hay, silage) is usually determined by quality standards, and accurate information on forage quality is a prerequisite for proper rationing and efficient balancing of stock feeding programs. Of course, the factors affecting the plant composition of a certain forage crop are similar whether the crop is irrigated or not. However, management decisions which control high forage quality, such as water and nutrient supply, are often more easily implemented when crops are grown under irrigation than without.

In this chapter an evaluation and short summary of the concepts of forage quality are presented. Detailed reviews of these and related topics have been written by Hacker and Minson (1981), Van Soest (1982), Wilson (1982), and Minson (1983, 1990).

Forage quality of individual crops is discussed in the respective chapters of Part Two.

4.2 CHEMICAL AND STRUCTURAL COMPOSITION

Forages, in common with all plants, are made up of variously modified cells which have two major components: cell contents and cell-wall constituents. The *cell-contents fraction* contains most of the organic acids, soluble carbohydrates, crude protein, fats and soluble ash, which are potentially 'completely digestible'. The *cell-wall fraction* (CWC) includes lignin, cutin and silica, which are classified as 'completely indigestible'. Much of the hemicellulose and cellulose of the CWC is protected from the rumen flora by a layer of indigestible lignin and these two constituents can be divided into potentially digestible and indigestible fractions (Fig. 4.1). Figure 4.1 also shows the major distribution of chemical fractions among the different types of plant tissues. The amounts of these fractions vary greatly among plant parts, species and development stage, and are digested in the rumen to varying extents. Resistance to rumen digestion was found to increase in the following order: mesophyll and phloem < epidermis and parenchyma sheath < sclerenchyma < lignified vascular tissue. Lignification — the presence of phenolic and other compounds in the cell walls — was found to be a major factor limiting rumen microbial degradation (Van Soest, 1967; Akin, 1982; Akin and Chesson, 1989). The differences in

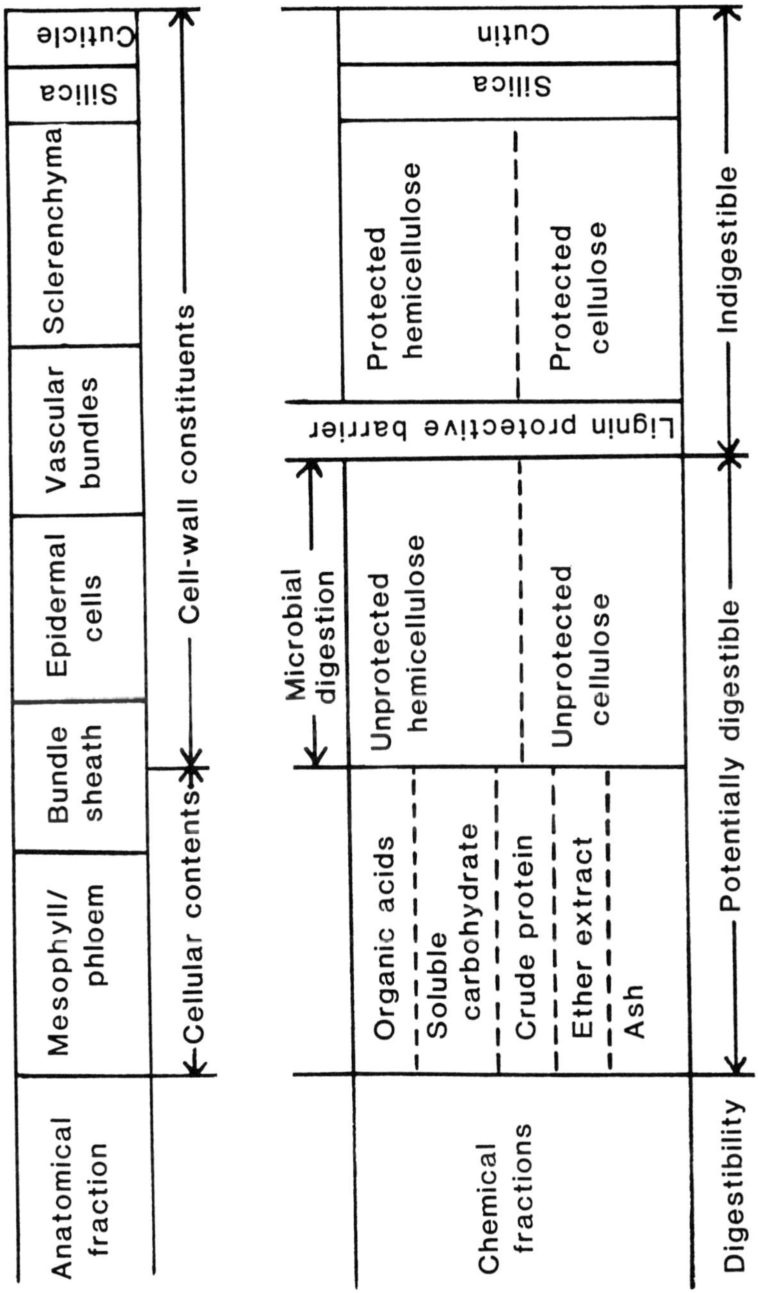

Fig. 4.1. Conceptual model of the relation between plant anatomy and chemical fractions indicating areas of potential digestibility. (From Minson, 1990).

structure lead to a wider range of digestibility of forages than of any other feed eaten by ruminants.

4.3 DIGESTIBLE ENERGY

The energy requirements of ruminants are expressed in terms of metabolizable energy (ME) or net energy (NE); the units used are MJ kg DM^{-1}. The ME value of a feed takes into account the energy lost in the urine and as methane, in addition to the fecal loss. The NE value of a feed is determined by subtracting from the ME value, the additional heat produced when a forage is eaten (Blaxter, 1989).

Since determining ME and NE values of forages is an elaborate procedure, most energy values are limited to measuring the loss of forage DM in the feces being expressed as *dry-matter digestibility* (DMD). DMD is expressed either in units of g kg^{-1} or as a coefficient:

$$\text{DMD} = \frac{\text{I} - \text{F}}{\text{I}} \times 1000$$

where I = the quantity of DM eaten and F = the quantity of DM excreted. The concept of digestibility is also applied to the *organic dry matter* (ODM) of forage, the ash content of which — although soluble but not containing energy — is typically 7%. It is therefore assumed that the *digestible organic matter in dry matter* (DOMD) is approximately 93%, and hence DOMD is about 5–6 units less than OMD. The *total digestible nutrient* (TDN) value is similar to that of DOMD. It is important not to confuse these expressions of digestibility! The relationship between ME and OMD is given by:

$$\text{ME (MJ per kg DM)} = \text{DOMD (\%)} \times 0.16$$

assuming the energy contribution of the DOM to be 16 MJ kg^{-1} (Murray, 1989).

The concept of digestibility may also be applied to other chemical fractions of the feed e.g. crude protein and crude fiber, and are designated: *digestible crude protein* (DCP) and *acid detergent fiber* (ADF), respectively.

4.4 VARIATION IN DIGESTIBILITY

Sources of the large variation in DMD values of forages are associated mainly with differences between species (sometimes also cultivars), plant part, growth stage, water and nutrient supply, season, and processing.

Temperate versus tropical species

The mean DMD coefficient of a great number of temperate grasses was found to be 0.13 higher than that of tropical grasses (Minson and McLeod, 1970).

The lower digestibility of tropical grasses is due to differences in anatomical structure associated with the different photosynthetic pathways and the higher temperature at which tropical grasses are normally grown. Leaf blades and stems of tropical grasses have more vascular bundles per unit cross-section area and hence more sites for lignification than temperate grasses (Hanna et al., 1973; Wilson, 1983).

Cultivars

Interspecific and varietal differences in DMD values of tropical grasses are usually small when they are harvested at a similar growth stage. The difference in DMD values between cultivars of corn is associated mostly with the proportion of ears in the forage. The incorporation of brown midrib mutant genes in lines of forage corn and sorghum yielded consistently higher DMD values because of less lignification of leaf and stem tissue. However, the yields are depressed and plants show increased sensitivity to lodging (Cherney et al., 1991).

Differences in DMD values between species and cultivars of temperate legumes also are rather small when the plants are harvested at similar growth stage. Breeding alfalfa with improved DMD, i.e., reduced lignin concentration, has yet not yielded commercial cultivars (Hill et al., 1988).

Plant parts

Only general trends of differences in DMD between plant parts are discussed here; the topic has been reviewed thoroughly by Hacker and Minson (1981). Of all plant parts, roots (e.g. fodder beet) and kernels show the smallest, and stems the largest, variation in DMD. The high DMD of roots and kernels is related to their function as storage organs and their high proportion of available carbohydrates.

Among grass species leaf DMD varies by up to 40 percentage units attributable mainly to age effects, temperature, nutrition and position of attachment. When plants or pasture are harvested at intervals, it is usual to find a steady decline with time in the DMD of the leaf fraction — more rapid in grasses than in legumes. This is explained in part by the abscission of older leaves in the legumes, whereas in the grasses there is less abscission. The decline in DMD of grasses with time is moderated by the addition of new leaves at rates which depend on growth stage and environmental factors. For example, Wilson and 't Mannetje (1978) found in certain tropical grasses that the daily decline in DMD of individual leaves exceeded values for entire leaf fractions. On the other hand, the rate of decrease of DMD of an individual leaf of corn is similar to that for the entire leaf fraction, due to lesser digestibility of later formed leaves (Deinum and Dirven, 1971).

Leaf DMD of legumes deteriorates much more slowly than stem DMD. In alfalfa, for example, leaf DMD varied only slightly with maturity and position within the canopy (Kilcher and Heinrichs, 1974). In contrast, the crude protein (CP) concentration of alfalfa leaves declined continuously from the vegetative to the early seed stages, with leaves low in the canopy having less CP than those near the canopy top (Buxton et al., 1985).

The variation in DMD of stems exceeds that of any other plant part, mainly due to aging effects. Young stem tissue is as, or even more digestible than leaf in grasses. However, with age and season, DMD of stems declines because of the increase in structural matter mostly associated with flowering. The DMD of stems (stalks) of silage corn at harvest is lower than any other part of the plant mainly because of the presence of highly lignified tissues in the rind of the stem. On the other hand, the pith of the stem can contain highly digestible carbohydrates, which moderate stem DMD (Struik, 1982).

In contrast to leaves, alfalfa stems show a decline in DMD with advancing maturity, and stem bases are less digestible than stem tops. Digestibility of stem segments is related closely to the lignin concentration of cell walls (Marten et al., 1988).

Soil moisture

In field studies, water-stressed alfalfa and grasses often had a similar or higher DMD than non-stressed plants, possibly due to a reduction in the stem/leaf ratio and higher DMD of the stressed leaves (Vough and Marten, 1971; Snaydon, 1972; Mislevy and Everett, 1981; Wilson, 1983).

4.5 EVALUATION OF FORAGE DIGESTIBILITY

Concentrations of CP, nitrogen-free extracts (mostly non-structural carbohydrates) and specific minerals in forage are of major interest since these are required nutrients for ruminants and their determination is often a routine procedure in the chemical analysis of forage. The Van Soest (1967) systems of fibrous feed analysis have become primary standards for chemical evaluation of forages: ADF as a predictor of digestibility and neutral detergent fiber (NDF) as a predictor of intake potential of forage.

The concentrations of most of the above fractions in individual crops are discussed in Part Two. Also moisture concentration is of major importance, since it indicates the conditions and type of forage curing and preservation. Moisture concentration must also be known in order to calculate nutrient concentration on a DM basis.

Although reasonable estimates of nutrients can be made, the rate and extent of their degradation in the rumen (i.e., digestibility), and their prediction, have become topics of intensive research in recent decades. Some of the currently used techniques for predicting forage digestibility are:

In vivo

Digestion trials with animals (usually with sheep) are carried out to measure digestible energy (DE) and ME in forages (Schneider and Flatt, 1975). These trials are time-consuming and require elaborate equipment. However, they are essential as standards against which the accuracy of indirect methods is assessed (Van der Honing and Steg, 1990).

Bioassay

The most common two-stage *in vitro* technique (Tilley and Terry, 1963) is considered the most precise laboratory technique for estimating digestibility of forage DM and OM. Provided the rumen donor animals are fed on feeds similar to those to be tested, and that control forages are run in each batch, estimates of *in vitro* digestibility of dry matter (IVDM) and of organic matter (IVOMD), are in good agreement with those of *in vivo* measurements.

The '*in sacco*' technique, whereby silk or nylon bags are incubated within the rumen, offers a compromise method which is technically easier than *in vivo* or *in vitro*, but demands great care to produce meaningful results (Orskov and Reid, 1989).

Physical measurement

Norris et al. (1976) showed that the *in vivo* DMD of forage can be predicted from the reflectance from finely-ground samples (of dried forages) of near-infrared radiation (NIR). Reflectance spectra of a large number of forages were amassed on a computer and a multiple regression model was used to identify analytically useful absorption bands which could predict both chemical attributes and animal response. For calibration of the scanning NIR spectrometer, a large set of extremely well analyzed materials from a wide distribution of samples is required, all typical of those to be tested in the future. Lately this new technology has become a practical tool used in extension programs, in private testing programs to aid buyers and sellers of hay, by forage breeders, and in research (Biston, 1989; Marten et al., 1989).

4.6 CONCLUSIONS

Forage quality, in the narrow sense, is determined by the plant's digestible energy and protein content. The concentrations of these nutrients depend on plant species, stage of growth, growing conditions, and harvesting and storage management. High digestible dry matter content of forages is associated with characteristics such as: high leaf-to-stem ratio, low amounts of cell-wall constituents, dough stage of kernels and increased non-structural carbohydrate content in stems and roots. Traditional conservation methods of field curing and ensiling decrease digestible dry matter compared with freshly harvested forage.

The digestible dry matter content of a feed is the proportional difference between the quantity of dry matter eaten and the quantity excreted in the feces. The concept of digestibility is closely related to other energy parameters such as: organic-matter digestibility, total digestible nutrients, and metabolizable energy. A wide array of biological (*'in vivo'* and *'in vitro'*), chemical and physical (near infrared reflectance spectroscopy (NIRS)) techniques are available for the evaluation of forage digestibility. The choice of depends mainly on the time available, precision required, accessibility and costs involved.

REFERENCES

Akin, D.E. 1982. Microbial breakdown of feed in the digestive tract. pp. 201–219. In: J.B. Hacker (Ed.) Nutritional Limits to Animal Production from Pastures. CAB, Farnham Royal, UK.

Akin, D.E. and Chesson, A. 1989. Lignification as the major factor limiting forage feeding value especially in warm conditions. pp. 1753–1760. Proc. 16th Int. Grassland. Congr. (Nice, France).

Biston, R. 1989. Present and future of NIR in the field of forages. pp. 1854–1856. Proc. 16th Int. Grassland Congr. (Nice, France).

Blaxter, K.L. 1989. Energy metabolism in animals and man. Cambridge Univ. Press, Cambridge, UK. pp. 336.

Buxton, D.R., Wedin, W.F. and Marten, G.C. 1985. Forage quality in stratified canopies of alfalfa, birdsfoot trefoil and red clover. Crop Sci. 25: 273–279.

Cherney, J.H., Cherney, D.J.R., Akin, D.E. and Axtell, J.D. 1991. Potential of brown-midrib low-lignin mutants for improving forage quality. Adv. Agric. 46: 157–198.

Deinum, B. and Dirven, J.G.P. 1971. Climate, nitrogen and grass. 4. The influence of age on chemical composition and *in vitro* digestibility of maize (*Zea mays* L.) and tall fescue (*Festuca arundinacea* Schreb.). Neth. J. Agr. Sci. 19: 264–272.
Hacker, J.B. and Minson, D.J. 1981. The digestibility of plant parts. Herbage Abstr. 51: 459–482.
Hanna, W.W., Monson, W.G. and Burton, G.W. 1973. Histological examination of fresh forage leaves after *in vitro* digestion. Crop Sci. 13: 98–102.
Hill, R.R., Shenk, J.S. and Barnes, R.F. 1988. Breeding for yield and quality. In: A.A. Hanson, D.K. Barnes and R.R. Hill (Eds.) Alfalfa and Alfalfa Improvement, ASA, CSSA and SSSA Monogr. 29: 809–825, Madison, WI.
Kilcher, M.R. and Heinrichs, D.H. 1974. Contribution of stems and leaves to the yield and nutrient level of irrigated alfalfa at different stages of development. Can. J. Plant Sci. 54: 739–742.
Marten, G.C. 1985. Factors influencing feed value and effective utilization of forage for animal production. Proc. 15th Int. Grassland Congr. (Kyoto, Japan). pp. 52–58.
Marten, G.C., Buxton, D.R. and Barnes, R.F. 1988. Feeding value (Forage quality). In: A.A. Hanson, D.K. Barnes and R.R. Hill (Eds.) Alfalfa and Alfalfa Improvement, ASA, CSSA and SSSA Monogr. 29: 464–491. Madison, WI.
Marten, G.C., Shenk, J.S. and Barton II, F.E. 1989. Near infrared reflectance spectroscopy (NIRS): Analysis of forage quality. U.S. Dep. Agric., Agric. Handbk. No. 643.
Minson, D.J. 1983. Forage quality: Assessing the plant-animal complex. Proc. 14th Int. Grassland Congr. (Lexington, KY). pp. 23–29.
Minson, D.J. 1990. Forage in Ruminant Nutrition. Academic Press Inc. New York, NY. pp. 483.
Minson, D.J. and McLeod, M.N. 1970. The digestibility of temperate and tropical grasses. Proc. 11th Int. Grassland Congr. (Surfers Paradise, Australia). pp. 719–722.
Mislevy, P. and Everett, P.H. 1981. Subtropical grass species response to different irrigation and harvest regimes. Agron. J. 73: 601–604.
Murray, I. 1989. Recent developments in the evaluation of ruminant feeds by chemical analysis and NIR spectroscopy. Proc. 3rd Egypt–Brit. Conf. Anim., Fish and Poultry Prod. (Alexandria, Egypt). pp. 33–41.
Norris, K.H., Barnes, R.F., Moore, J.E. and Shenk, J.S. 1976. Predicting forage quality by infrared reflectance spectroscopy. J. Anim. Sci. 43: 889–897.
Orskov, E.R. and Reid, G.W. 1989. Comparison of chemical and biological methods for predicting feed intakes and animal performance. pp. 158–66. In: E. Chesson and E.R. Orskov (Eds.) Physico-Chemical Characterization of Plant Residues for Industrial and Feed Use. Elsevier Sci. Publ., Amsterdam, the Netherlands.
Schneider, B.H. and Flatt, W.P. 1975. The evaluation of feeds through digestibility experiments. Univ. Georgia Press, Athens, GE. pp. 423.
Snaydon, R.W. 1972. The effect of total water supply and of frequency of application upon lucerne. Aust. J. Agr. Res. 23: 253–256.
Struik, P.C. 1982. Production pattern, chemical composition and digestibility of forage maize (*Zea mays* L.). Med. Vakgr. Landb. Grassl. 64., Landbouwhogeschool, Wageningen, the Netherlands. pp. 28.
Tilley, J.M.A. and Terry, R.A. 1963. A two-stage technique for the "*in vivo*" digestion of forage crops. J. Brit. Grassland Soc. 18: 104–111.

Van der Honing, Y. and Steg, A. 1990. Comparison of energy evaluation systems of feeds for ruminants. pp. 1-19. In: J. Wiseman and D.J.A. Cole (Eds.) Feedstuff Evaluation, Butterworth, London, UK.

Van Soest, P.J. 1967. Development of a comprehensive system of feed analyses and its application to forages. J. Anim. Sci. 26: 119–128.

Van Soest, P.J. 1982. Nutritional Ecology of the Ruminant. O & B Books Inc., Corvallis, OR, 374 pp.

Vough, L.R. and Marten, G.C. 1971. Influence of soil moisture and ambient temperature on yield and quality of alfalfa forage. Agron. J. 63: 40–42.

Wilson, J.R. 1982. Environmental and nutritional factors affecting herbage quality. pp. 111–113. In: J.B. Hacker (Ed.) Nutritional Limits to Animal Production from Pastures. CAB, Farnham Royal, UK.

Wilson, J.R. 1983. Effects of water stress on *in vitro* dry matter digestibility and chemical composition of herbage of tropical pasture species. Aust. J. Agr. Res. 34: 377–390.

Wilson, J.R. and 't Mannetje, L. 1978. Senescence, digestibility and carbohydrate content of buffel grass and green panic leaves in swards. Austr. J. Agr. Res. 29: 503–516.

Part Two:
The Production

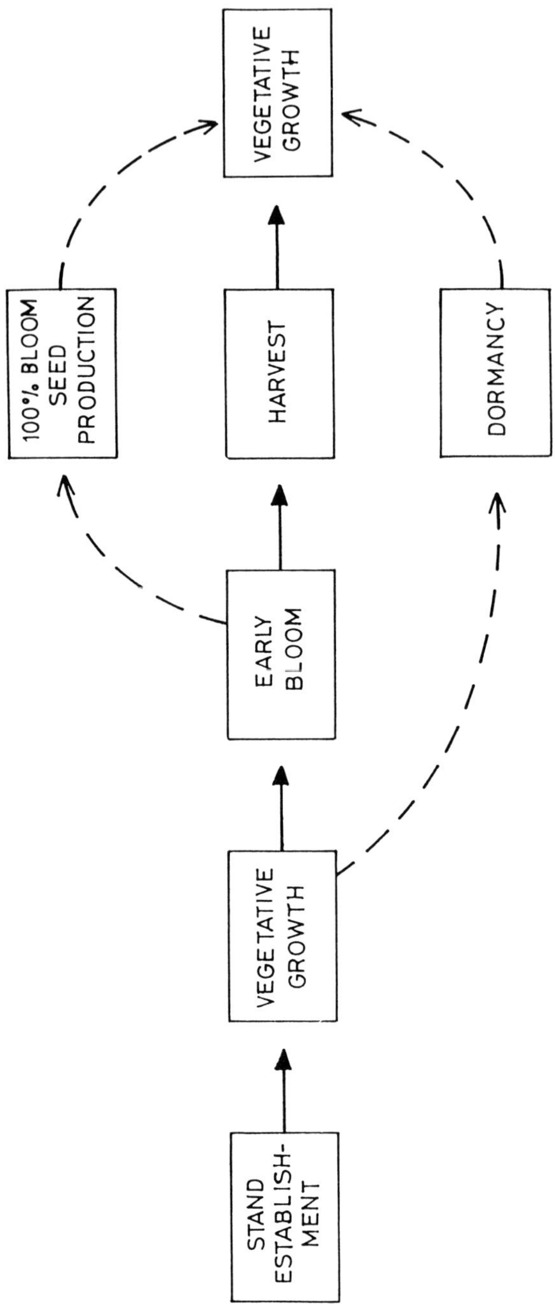

Fig. 5.1. Alfalfa growth and development cycle. (From Anon., 1985).

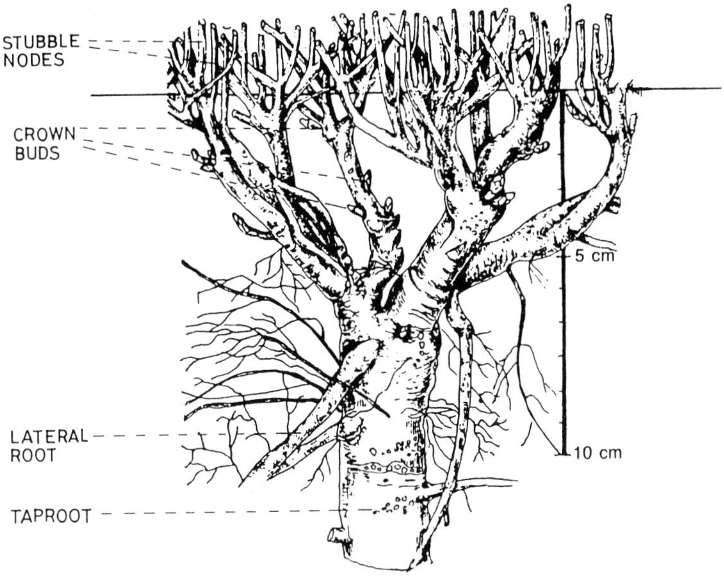

Fig. 5.2. Schematic presentation of a mature crown of alfalfa. (Adapted from Mokeyeva, 1940).

5.2.1 Canopy Structure

Three stages of canopy development are distinguished: (i) shoot initiation, (ii) vegetative growth, and (iii) reproductive growth. Potential shoot density (i.e. number of shoots per m^2) is determined during the shoot initiation stage, whereas shoot size, shoot quality and realized stem density (i.e. number of stems per m^2 ground area) are determined by the vegetative and reproductive growth stages.

Shoot initiation

Shoots initiate from stubble nodes in subsequent regrowth after cutting and from crown buds after a period of dormancy. The location of these buds is shown in Figure 5.2. The number of shoots per plant increases with age and in any growth cycle the maximum is usually set within 14 d of the start of regrowth and then declines as the canopy matures (Leach, 1969). The yield of regrowth depends primarily on the number of shoots and the time when each shoot resumes growth. Shoot initiation in the hot summer is considerably less than in the spring (Marble, 1989). The reduction of soil water substantially reduces the number of shoots which regrow after defoliation (Cohen et al., 1972; Perry and Larson, 1974). Brown and nean region, the relatively high summer temperatures are conducive to early flowering and to short growth intervals. The number of harvests from spring to autumn can reach nine or more. During the mild winters prevailing in these regions, alfalfa is usually harvested at long intervals. The few harvests during the autumn and winter seasons barely affect the yield level of the spring harvests. Continuous harvesting of alfalfa throughout the autumn and winter months is expected to increase total seasonal yield of high quality hay by at least 5–10% per year, depending on growing conditions (Worker and Marble, 1987).

5.3 ESTABLISHMENT

5.3.1 Soil Requirements

Soil characteristics favoring high production of irrigated alfalfa are: (i) reasonable depth (>1 m), (ii) good drainage and medium texture, (iii) absence of hardpan, (iv) slight alkalinity, and (v) relative freedom from salts (Marble, 1989). Shallow soils restrict root growth and therefore plants suffer sooner from water deficits, thus necessitating more frequent irrigation. Heavy soils are usually associated with low permeability and compaction which induce water logging, scald and root disease (Leath et al., 1988), unless proper water management is practiced. This involves: (i) deep tilling prior to seeding in order to break up compacted layers and leveling the land to eliminate spots in which water can accumulate, (ii) provision of adequate irrigation drainage outlets, and (iii) application of gypsum or sulfuric acid to increase soil permeability (Whitfield et al., 1986).

5.3.2 Seeding

Tillage and seedbed preparation

Plowing at a depth greater than 25 cm is usually not needed for the preparation of a satisfactory seedbed, depending on soil type, the amount of crop residue, and the degree of compaction as a result of the traffic of field equipment. Disking and harrowing generally suffice when alfalfa follows other crops in a regular crop rotation. In desert soils — particularly stratified fine sandy-loam soils and silty clay soils (Kaddah, 1977; Marble, 1989), or in slowly permeable clay-loam soils (Eck et al., 1977), soil profile modification by thorough mixing of the soil or by deep tillage led to significant increases in yield and in water use efficiency (WUE). Lower yields and lower WUE on unmodified soil resulted principally from inadequate penetration and the loss of water to evaporation, and from the effects of surface ponding of water after irrigation (Meek, 1978). After the soil was ameliorated with gypsum and/or by deep ripping, sprinkler irrigation was found to be more effective than flood irrigation in increasing the depth of wetting (Taylor and Olsson, 1987). A heavy-duty moldboard plow, chisel or slip-plow can be used for deep tillage; deep moldboard plowing is the most expensive of the three methods.

Alfalfa root development was reduced in irrigated sandy soils with low organic matter content and low bulk density. The addition of manure and fertilizer under deep plowing compared with disking, considerably increased yields in the seeding year (Meyer, 1971). Considering the relatively small size of the seeds, the preparation of a firm seedbed free of large, hard pieces of soil and favoring good soil-seed contact is essential for satisfactory seedling establishment.

Inoculation of seed

In areas where irrigated alfalfa is growing successfully, the usefulness of inoculating the seed with *Rhizobium meliloti* bacteria to ensure effective N fixation, has not yet been demonstrated unequivocally. Dust and irrigation water are apparently very useful carriers of these bacteria and generate the natural distribution of inoculant. However, in land recently brought under cultivation, or in soils with pH of 6.2 or less, seed inoculation is recommended in order to ensure adequate root nodulation for effective N fixation (Marble, 1989).

Date of seeding

Optimum soil temperature for seedling development was found to be in the range 21–24°C depending on the dormancy classification of the cultivar. This classification is based on the differential autumn-growth response of cultivars to temperature and daylength (Section 3.1.2). Seedlings under short photoperiods produce more crown buds and larger root systems — which improve winter survival — than under long days. By matching the optimum temperature and photoperiod for a cultivar with the corresponding seasonal conditions at a specific location, predictions are made regarding the most desirable time to seed (Teuber et al., 1987).

The advantages of seeding in autumn compared with the spring are an expected higher first-year yield, increased seedling vigor enabling better competition with winter weeds, and the development of a more vigorous and deeper root system (Marble, 1989). However, local conditions may dictate a more appropriate time of seeding, depending on the incidence of hot, dry winds, or a soil too wet to be worked, or the probability of harmful weeds, insects or birds during the autumn season.

Seeding rates and seeding

Seeding rates of 25–30 kg ha^{-1} of high-quality seed are recommended under irrigated conditions in California. The rates may be lower on extremely well-prepared seedbeds and when sprinkler irrigation is available. Higher seeding rates are required when seed is of low quality, seedbeds are poorly prepared, or in delayed autumn sowings — to ensure the survival of sufficient seedlings after the first winter season (Taylor, 1987). Heavy seeding tends to ensure better stands in areas with salinity problems (Lehman, 1979). Tesar and Marble (1988) summarized that 150 to 250 plants per m^2 are necessary for maximum yields in the year after seeding. Forty to 60 plants per m^2 may be adequate for maximum yield in older stands, provided the plants are uniformly distributed.

The use of a corrugated seeder roller effects shallow seeding and desirable compaction of the soil, except on sandy soil where seeding should be deeper. Compaction with a press wheel when seeding with a drill, or with a ring-roller after a broadcast seeding, will be useful to improve germination, especially during warmer periods and when moisture is limiting. Partial emergence of seedlings is often the result of too deep seeding in fluffy and powdery seedbeds. The use of a Brillion Seeder is recommended, since it has a roller in front to firm the soil and a roller following behind the seed drop which presses the seed into the soil at the optimum depth (Anon., 1971; Tesar and Marble, 1988). Many experiments support current seeding recommendations of shallow depths of 0.6–1.9 cm on most fine-textured soils and of 1.3–3.9 cm on coarse-textured soils when moisture conditions are less favorable (Tesar and Marble, 1988; Marble, 1989).

Causes for establishment failure

Tesar and Marble (1988) list the following causes for failure of a viable seed to produce a plant: (i) poor seedbed preparation, (ii) seeding too deep, (iii) inadequate moisture after germination, (iv) freezing, (v) disease, (vi) insects, (vii) competition for light and nutrients (with other alfalfa plants or weeds) and (viii) damage caused by herbicide and/or insecticides.

Weed control

Excessive amounts of weeds in seedling alfalfa can result in total or severe stand loss before cutting. It may be advisable to adjust the date of planting in order to avoid weed infestation in a field that has a history of severe weed problems. Alternatively, a preplant irrigation far enough in advance of planting to permit germination of any weed seeds present in the field, followed by shallow cultivation(s), may be useful. The use of a pre-emergence herbicide is also recommended. Cultural and chemical control methods of weeds in alfalfa seedling fields are reviewed by Peters and Linscott (1988) and Norris (1980).

Companion crops

In most Near Eastern countries with mild to warm winters, the optimum time to seed is at the beginning of the autumn season. In desert areas with high winds and sand storms, alfalfa should be seeded with a cereal grain as a companion crop for the protection of the young seedlings. Marble (1989) recommends for that purpose the use of oat seeded before the alfalfa, or a wheat–alfalfa mixture. Severe competition from the cereal crop can be avoided by pasturing, green chopping, or making hay or silage rather than allowing the grain to mature.

The seeding of oat with alfalfa usually increases the forage yields of the first cutting and reduces competition from weeds (Lanini et al., 1991). Alfalfa–grass mixtures are not recommended since hays fetch lower market prices.

5.3.3 Irrigation

Alfalfa may be seeded in a damp seedbed or in dry soil. A preseeding irrigation has the following advantages: (i) it saturates the soil profile, (ii) firms the seedbed, (iii) stores moisture and thereby reduces water demand when water is needed by annual crops, (iv) improves nodulation, and (v) germinates weeds prior to seeding thus reducing possible weed contamination of the new stand (Marble, 1972). For the last reason, preseeding irrigation is usually given when sprinklers are available for irrigation. Subsequent light irrigations may be needed to replenish moisture deficits in the top soil layer for the support of initial root and seedling growth.

On heavy dry soils which are flood irrigated, the practice of preirrigation is generally unsuccessful (Lehman, 1979). For successful germination of a stand of alfalfa seeded in a mulch, seeds must be placed fairly deep. If there are dry areas in the soil, no seed will germinate there. Subsequent irrigation of deep-placed seeds will prevent seedling emergence and spotty stands will be the result. The practice of withholding irrigation following emergence for the purpose of increasing root penetration is invalid, because seedling root growth is suppressed more than shoot growth by moisture stress (Sheaffer et al., 1988).

On dry seedbeds shallow seeding is required and flooding is needed to provide soil moisture. This first irrigation should be gentle, penetrate deep, and cover the soil as completely as possible. Care is taken to prevent washing of soil, which causes excessive loss of seed. Additional light irrigations are necessary to soften a soil crust when it forms after irrigation, in order to enable seedlings to emerge.

Robinson et al. (1968) reported that the use of sprinkler irrigation to germinate alfalfa has proven superior to flood irrigation, in particular when seeding in autumn on saline soils.

5.3.4 Fertilization

Fertilizer recommendations based on soil tests developed for the area where alfalfa is to be grown, will guide the grower in deciding whether or not to incorporate fertilizers prior to seeding. Tentative soil analysis values for possible yield response of alfalfa to P, K and S fertilizer in California are presented in Table 5.2.

TABLE 5.2

Guide for interpretation of results from soil analyses for possible yield response of alfalfa to P, K and S fertilizer application prior to seeding in California. (From Reisenauer et al., 1978)

	Bicarbonate P ($mg\ kg^{-1}$)	Exchangeable K ($mg\ kg^{-1}$)	LiCl–Soluble S ($mg\ kg^{-1}$)
Response likely	< 6	<50	< 5
Response not likely	>10	>80	>10

The pH of most soils in semi-arid and arid regions is in the range 6.0–7.5 and is recommended for good alfalfa production (Woodruff, 1967). Soil pH values less than 5.5 are usually associated with poor alfalfa growth and deficient nodulation. Applications of either lime or fertilizer N may remedy these deficiencies (Lanyon and Griffith, 1988).

Phosphorus is usually the most important limiting nutrient and is applied as basic fertilizer prior to seeding. In loam or clay–loam desert soils, other nutrients are rarely needed, although occasionally S, K, B, or Mg may limit production (Marble, 1989). The application of fertilizers on established alfalfa stands is dealt with in Section 5.5.

5.3.5 Date of the First Harvest

New alfalfa stands usually are not harvested before seedlings have developed two to four stems; cutting at an earlier stage is likely to produce less vigorous plants. Harvesting when the soil is still wet should be avoided. In sandy and sandy-loam soils, the alfalfa taproot should be at least 35 cm below the soil surface before the first mechanical harvest takes place. Plants growing in light-textured soil with small root systems at the time of first harvest, were generally weakened and few survived the double blow of root-restricting soil compaction and mechanical damage to regrowth (Grimes et al., 1978; Sheesley, 1982; Sheesley et al., 1982).

5.4 CROP WATER REQUIREMENTS AND IRRIGATION

The alfalfa crop has a relatively high water demand and long production season, and is exposed to periodic harvesting. These and additional factors require due consideration in attempts to improve water conservation practices. Reviews of alfalfa irrigation have been written by Jensen (1980), Finkel (1982) and Guitjens (1990); and of soil–water–plant relations by Doorenbos and Kassam (1979), Heichel (1983) and Sheaffer et al. (1988). Irrigation management for alfalfa in California was described by Hanson (1991).

5.4.1 Water-Production Functions

Transpiration (T)

A well-watered alfalfa crop transpires maximally when its foliage covers the ground completely. Little evidence was found that the plants exert stomatal control over water loss until the soil is depleted to about half of the available soil moisture (Van Bavel, 1967; Ludlow, 1980; Heichel, 1983). This insufficient stomatal regulation with increasing soil water deficits is often considered a major cause for the relatively high consumptive water use of alfalfa compared with crops like berseem, ryegrass, wheat and fodder beet. Cuticular transpiration has been suggested as an additional source of leaf water loss since the cuticle on the epidermis of the alfalfa leaf was observed to be very thin or virtually absent (Versteeg, 1985). Besides high stomatal conductance, other factors, such as small leaves with high boundary conductances, and high stem and root densities, contribute to high transpiration rates (Sheaffer et al., 1988). The transpiration coefficient (TRC) is usually somewhat lower in summer than in winter. Values of TRC differ sometimes between cultivars and genotypes (Briggs and Shantz, 1913; Pfeiffer et al., 1980; Sheehy et al., 1980; Valdivia and Zipori, 1982; Hattendorf et al., 1990). The selection of genotypes with lower TRC values could be instrumental in improving the water use efficiency (WUE) of alfalfa.

Evapotranspiration (ET)

After harvesting, T is negligible because of the low number of leaves in the stubble. At the same time, evaporation from the soil is high, especially after irrigation, because the soil is not protected by a canopy of leaves. Conversely, with foliage regrowth after cutting and the increase of LAI, T increases as a fraction of full cover in proportion to the increase in radiation intercepted by the foliage. At that time, evaporation from the soil surface becomes minimal. ET is a function of the time of year, geographic location and growth stage. The higher the ET, the greater the yield until maximum ET is reached. Maximum ET of well-watered alfalfa is usually reached at full plant development which ordinarily is associated with LAI of 3 (Stegman et al., 1980). At that stage of development the rate of T is similar to that of ET. Evapotranspiration from an alfalfa stand can vary considerably during the season, between growing cycles and between years, because of variation in evaporative demand and canopy structure. Smeal et al. (1991) developed a multiple regression equation in which forage yield was described as a function of T, growing degree day (GDD) accumulation, average daily solar radiation, year and harvest number within year. Total seasonal yield was found to be a function of T and year, which together explained 90% of yield variability.

The *crop coefficient* (K_c) for alfalfa is not a constant, since periodic harvests result in extended periods of small cover. K_c varies from a low value of 0.4–0.5 directly after cutting, to a peak of 1.15–1.25 with full cover, according to climatic factors such as wind and humidity (Doorenbos and Pruitt, 1977). This variation in K_c should account for the change in water consumption with stage of development during each growth cycle. In many experiments, using line-source sprinklers or lysimeters, it was shown that yield is often a linear function of ET. An example of linearity is presented in Figure 5.4. The slope of the regression line shown in this figure defines the WUE. The slope varies with factors that affect ET and crop management (e.g. cultivar, stand, harvest date, disease level, nutrient supply). The WUE of alfalfa in different environments is shown in Table 5.3. Higher WUE values are usually observed in spring and autumn than in late summer, when ET is considerable and growth is poor. The simulated WUE of approximately 1.9 kg DM m^{-3} of alfalfa from Davis was approximately half that of grain maize but not very different from silage maize (Loomis and Wallinga, 1991). It should be noted however, that the usefulness of WUE as a water-production function is limited since daily and seasonal water use are influenced markedly by climate and method of water application.

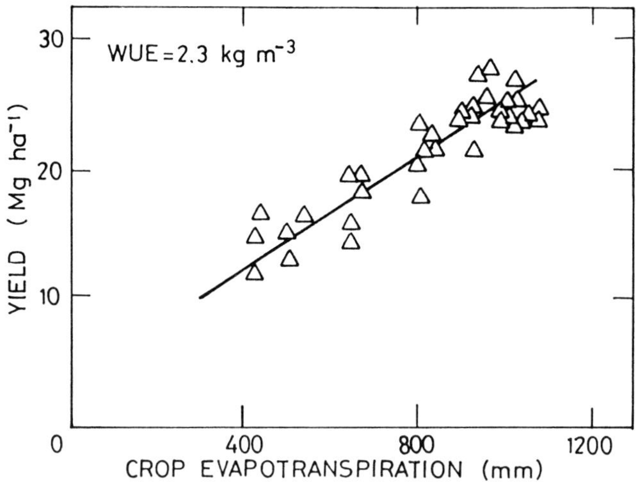

Fig. 5.4. Annual dry matter production by irrigated alfalfa as a function of crop evapotranspiration. San Joaquin Valley, Calif. (Adapted from Grimes et al., 1992).

TABLE 5.3

Average water use efficiency (WUE) of irrigated alfalfa in different environments. (Adapted from Loomis and Wallinga, 1991)

Location	WUE (kg m^{-3})	Reference
Reno, Nevada	1.5	Guitjens (1990)
North Dakota	1.6	Bauder et al. (1978)
Lubbock, Texas	1.7	Bolger and Matches (1990)
Cyprus	1.4–1.8	Metochis and Orphanos (1981)
Logan, Utah	1.4–2.2	Retta and Hanks (1980)
Parlier, Calif.	2.3	Grimes et al. (1992)
Davis, Calif.	1.9*	Denison and Loomis (1989)
Imperial Valley, Calif.	2.0*	idem
San Camilo, Peru	1.5	Zipori and Valdivia (1982)

*Simulated.

5.4.2 Sensitivity to Water Stress

Root growth

Quantitative information on root growth and development of alfalfa under field conditions is scarce because of the difficulties involved in accurately sampling the entire root system. Rates of root proliferation therefore are mostly inferred from the patterns of soil water extraction at various depths (Heichel, 1983).

The root system of alfalfa plants is strongly dominated by a tap root, which usually branches into secondary roots in the soil profile, wherever water is available (Abdul-Jabbar et al., 1984). The root system is further characterized by an extensive arrangement of fibrous roots which was found to decline exponentially with soil depth (Meyerhoff, 1984). Fibrous roots proliferating in the upper 20 cm of the soil profile bear the largest number of nodules. Tap-root mass and the length density of fibrous roots seem to fluctuate over time of a regrowth cycle and also seasonally, and are apparently related to changes in the content of nonstructural carbohydrates in the roots (Feltner and Massengale, 1965; Robison and Massengale, 1968; Heichel, 1983; Denison and Loomis, 1989).

Moisture stress

Alfalfa plants are water-stressed when the available soil moisture cannot satisfy the evaporative demand of the atmosphere; this results in a deficit ET and a reduction in DM. As a result of water stress, stems elongate less and leaves remain smaller (Bauder et al., 1978; Brown and Tanner, 1983; Sharratt et al., 1983). Both the growth rate (GR) and relative growth rate (RGR) of stressed leaves and stems were significantly lower than well-watered leaves and stems. Development of water stress in the second half of the growth cycle reduced leaf size and internode length, but not numbers of leaves and internodes or stem density. The percentage of total forage dry weight in leaves decreased linearly with stem height, resulting in a better quality forage in stress-stunted foliage. However, stem density decreased significantly when the stress occurred during the first 14 d of regrowth (Brown and Tanner, 1983). These findings draw attention to the difference in growth response to moisture stress during specific growth stages. Such information is essential when opting for optimal yield in a situation with limited water supply (Guitjens, 1990).

5.4.3 Criteria for Irrigation Scheduling

When and how much to irrigate are crucial decisions in alfalfa irrigation management. Since an inadequate supply of readily available soil moisture reduces ET and DM production, irrigation should take place before this situation develops. If winter precipitation has not been sufficient to recharge empty soil moisture storage in the root zone, an alfalfa stand should be irrigated as soon as possible around the time the plants are breaking dormancy.

Available soil moisture depends on soil physical properties and the depth of the soil profile. Heavy clay soils with slow infiltration are prone to water logging, but may be improved through the application of gypsum (Whitfield et al., 1986). The yield of alfalfa increases with drainage in a fine-textured soil (Bornstein et al., 1984). In shallow soils there is less volume for root development; plants will thus experience moisture stress sooner and therefore require more frequent irrigation. Deep ripping in desert soils increased water infiltration and alfalfa yields (Versteeg, 1982). The amount of water application depends on depleted soil moisture in the root zone and on the irrigation efficiency. Abdul-Jabbar et el. (1984) reported that with line-source sprinkler irrigation, greatest total root mass, rooting depth and root-length distribution of alfalfa in a clay–loam soil occurred at the highest irrigation rate (Fig. 5.5). The maximum root mass was correlated with an average soil depth of 1.6 m. These authors found that, excluding the tap root, root mass was correlated linearly with yield and curvilinearly with ET. Nearly two thirds of the root mass was concentrated in the top 45 cm of soil. Guitjens (1982) developed a model that estimates the relative uptake of soil water with depth in an extraction pattern of 40-30-20-10% for four equal increments of the total root depth. This represents a 70% extraction from the upper half of the root zone. Similar extraction patterns are reported by Bula and Massengale (1972). Root depths depend on any impedance to root development in the soil profile and moisture availability.

Allowable depletion of soil moisture (i.e., the percent available soil moisture that can be used without suppressing crop yield) is usually

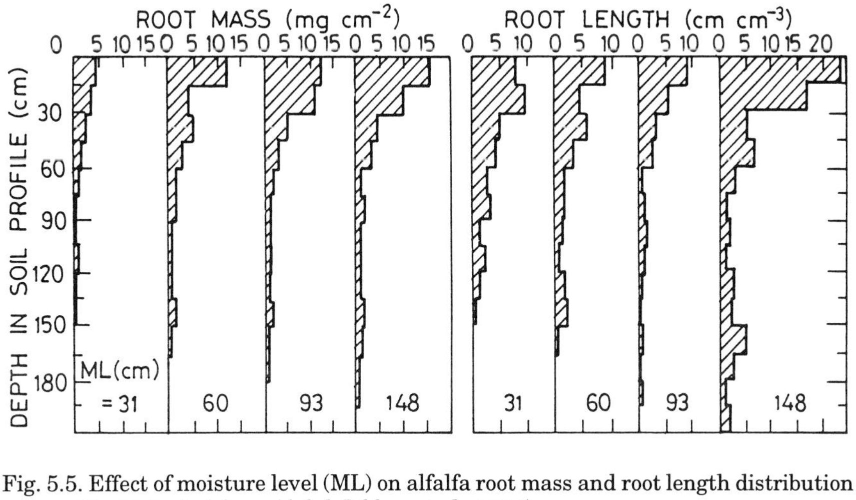

Fig. 5.5. Effect of moisture level (ML) on alfalfa root mass and root length distribution in the soil profile. (Adapted from Abdul-Jabbar et al., 1984).

approximately 50% for alfalfa (Hanson, 1991). The number of days between irrigations therefore depends on the allowable soil moisture depletion and differs with soil type and time of the year. If two-thirds of the root mass is concentrated in the top 45 cm of soil, irrigation should supply adequate amounts of water to this soil depth (Guitjens, 1990).

Canopy regrowth after cutting is strongly dependent on soil moisture availability. When irrigation was deferred by 10 d, regrowth of alfalfa after cutting (in soil close to the wilting point) was reduced by almost by one half compared with irrigation applied immediately after cutting. Withholding irrigation by an additional 13 d reduced by two-thirds the rate of daily growth and the plants failed to attain the regrowth rates of plants irrigated sooner after cutting (Cohen et al., 1972). Similar results are reported by Hall and Eckert (1991). Unfortunately, delay of irrigation after harvesting for hay is often inevitable, since the hay bales cannot be removed from the field until the cut forage has cured, which requires a couple of days.

Limited water supplies

Manipulating water application schedules when water supplies are restricted is feasible, thanks to the outstanding drought resistance of the mature alfalfa plant. This characteristic is due mainly to the plants osmotic adjustment which contributes to continued assimilation (and preferential translocation of assimilates to crowns and roots) despite increasing soil water deficits (Wright et al., 1986).

Drought-imposed alfalfa also quickly recovers after irrigation is restored. Recent experiments carried out in California (Frate et al., 1991) and in Nevada (Guitjens, 1990) showed that two or three consecutive years without irrigation during the last one or two growth cycles did not reduce yield in the following year when full irrigation was resumed. An average of 1:0.8 (% saving irrigation water) : (% yield loss) for two consecutive years was found in Israel in various drought treatments lasting from three to six months during summer, compared with full irrigation (Kipnis et al., 1989). The drought stress treatments had no carry-over effects on regrowth during the following winter and early spring of the following year. Similar results are reported from Cyprus (Metochis and Orphanos, 1981).

The following recommendations for irrigation with limited water supplies in California were formulated by Hanson (1991):

(i) Irrigate between cuttings if the water supply is sufficient. However, regardless of supply, irrigate the early cuttings adequately. Any reduction in the number of irrigations per cutting should occur with the last cutting, then the next-to-last, and so forth; and

(ii) if the water supply is insufficient to apply at least one irrigation per cutting, irrigations should be eliminated starting with the last cutting, then the next-to-last cutting, and so forth.

5.4.4 Sensitivity to Low Water Quality

Salinity

Alfalfa plants are classified as moderately sensitive to salinity with a threshold EC level of 2 dS m^{-1}, a 50% yield reduction at 8.9 dS m^{-1}, and a yield decline to zero at EC = 16 dS m^{-1} (Bresler et al., 1982; Maas, 1987). Assadian and Miyamoto (1987) concluded from seedling emergence studies conducted in a greenhouse that alfalfa can be established with minimal damage from saline water with ca. 4 dS m^{-1} when sown at a depth of ca. 10 mm.

Bernstein and Francois (1973), using lysimeters, showed that alfalfa can tolerate high salinity levels in the lower portion of the root zone without significant yield reduction, provided that the upper portion of the root zone is maintained relatively free of salts. Plants apparently compensate for the reduced water uptake from the highly saline zone by increasing water uptake from the zone low in salinity. In a later study, Francois (1981) — again using lysimeters — investigated the question of how much salt can be stored by leaching in the lower root zone before it encroaches on the upper root zone, with consequent yield reduction. The general conclusion from that study was that irrigating to leach accumulated salt out of the upper root zone of alfalfa may be necessary, but that it can be accomplished during the winter or off-season, when water is more readily available. Shallower soil profiles may require more frequent leaching.

Sewage water

Experiments on subsurface drip irrigation with treated municipal waste water were conducted during a number of years in the region of Be'er-Sheva, Israel (Oron et al., 1989). A maximum yield of approximately 25 Mg DM ha^{-1} $year^{-1}$ was obtained from an irrigation system with lateral spacing of 1.0 m and emitter depth of 40 cm. The yield of sprinkler-irrigated alfalfa was slightly less. The level of nutrients in the waste water rendered unnecessary the application of fertilizer. Subsurface drip irrigation improved weed control, extended alfalfa survival, and enabled quicker renewal of irrigation after harvesting than sprinkle-irrigated alfalfa. The cited report does not mention whether subsurface irrigated plots required less water than sprinkler-irrigated plots.

Experiments carried out at Buckeye, Arizona showed that alfalfa hay yields were higher, but IVDMD values similar, from flood-irrigated plots with municipal waste water and pump water (50:50) — without added fertilizer — compared with plots receiving pump water (from local wells) alone — supplemented with P fertilizer (Day et al., 1982).

5.5 MINERAL NUTRITION AND FERTILIZER USE

5.5.1 Nutrient Availability in the Soil

Results from soil tests can serve as a good first approximation of the potential of available nutrients in the soil. However, additional factors, such as soil composition, nutrient distribution in the soil profile, moisture supply (Simpson and Lipsett, 1973; Heichel, 1983), potential rooting volume and successful exploitation of that volume by alfalfa roots, affect the quantity and rate of supply of nutrients to plants (Lanyon and Griffith, 1988). Also the amounts of fertilizer needed for the maintenance of available soil nutrient levels differ among soils, due to the possible nutrient release and depletion by leaching or the adsorption or fixation of ions onto clay or organic matter particles, especially of P and K. Thus, prediction of yield response following the application of fertilizer requires a great deal of complementary field experimentation.

Fixen and Ludwick (1983) and Havlin et al. (1984) conducted long-term field experiments with irrigated alfalfa on typical calcareous soils of contrasting composition in Colorado. Soil tests indicated that the clay-loam soil was marginal in available P and very high in K, whereas the loam soil was high in P but marginal in exchangeable K. Soil-test P in control plots declined over six production years to levels where significant yield responses from fertilizer P were expected. Soil-test K remained at high levels in both soils and no yield response to fertilizer K was observed. It was concluded that P and K fertilization of these soils would not be recommended unless improved forage quality justified the costs of fertilizer application.

From initial tests of a sandy soil in Kansas, it was calculated that the amounts of available P and K were insufficient for the growth and maintenance of optimum production of irrigated alfalfa (Ball and Ten Eyck, 1980). Soil-test P levels after 3 years decreased almost 40% in plots not receiving P fertilizer, whereas annual broadcast applications of P increased both yields and P levels in the forage. Even though the initial soil test results showed that alfalfa was expected to respond to fertilization with K, there has been no yield response to applied K, apparently a result of the release of K from soil minerals with time.

Levin et al. (1969) studied the yield response of irrigated alfalfa following the application of P fertilizer, either surface-disked or deep plowed, on an alluvial loam-clay soil in northern Israel. The initial available soil-P level (Olsen, 1954) was within the range of an expected low response to fertilization with P, and no yield response to the application of P fertilizer was obtained during the 3 years of the experiment.

In adjacent plots at the same location, similarly designed experiments were carried out with K fertilizer. The initial available soil K level was considered sufficient with respect to the annual fertilizer requirement (Woodruff, 1955; Dovrat, 1962; Hagin and Dovrat, 1963). No yield response to added potash was obtained in the first year of the experiment, but there was a (non-significant) yield increase in the second year and a greater than 50% higher yield during the third year, compared with the control. The change in available K in the top to 0–20 cm soil layer with time is shown in Figure 5.6. The decline in available K of the control in the first year was not accompanied by a decrease in forage yield but further decrease in soil K availability caused considerable yield loss. The deep application of K fertilizer followed by top-dressing resulted in a uniform and high level of available K in the surface-soil layer.

Kafkafi et al. (1977) found a steady decrease of the concentration of K in the top 0–20 cm soil layer of a sandy-loam soil in Israel during one growing season of irrigated alfalfa, accompanied by a decrease in forage yields and concentration of K in plants (Fig. 5.7). Fertilization with K markedly increased DM matter yields and stand persistence.

5.5.2 Nutrient Uptake by the Crop

The estimated nutrient removal by a healthy alfalfa crop harvested annually under California conditions is shown in Table 5.4.

Much larger amounts may be taken up as luxury uptake from very rich soils, or lower amounts are removed if the soils are deficient (Martin and Luckhart, 1971). These data show that relatively large amounts of N, K and Ca are taken up by the alfalfa crop, less of Mg, P and Fe, and only minute amounts of Cl, Mn, B, Zn, Co and Mo. Nutrients most commonly in low supply are K, P, S and B, and these will have to be supplied in fertilizer form. For further details with respect to the individual elements, the reader is referred to reviews by Tesar and Marble (1988) and Lanyon and Griffith (1988).

Analyses of alfalfa tissues provide an effective guide for evaluating the nutritional status of plants and are useful in appraising fertilization programs. Manuals for the interpretation of plant analysis of alfalfa,

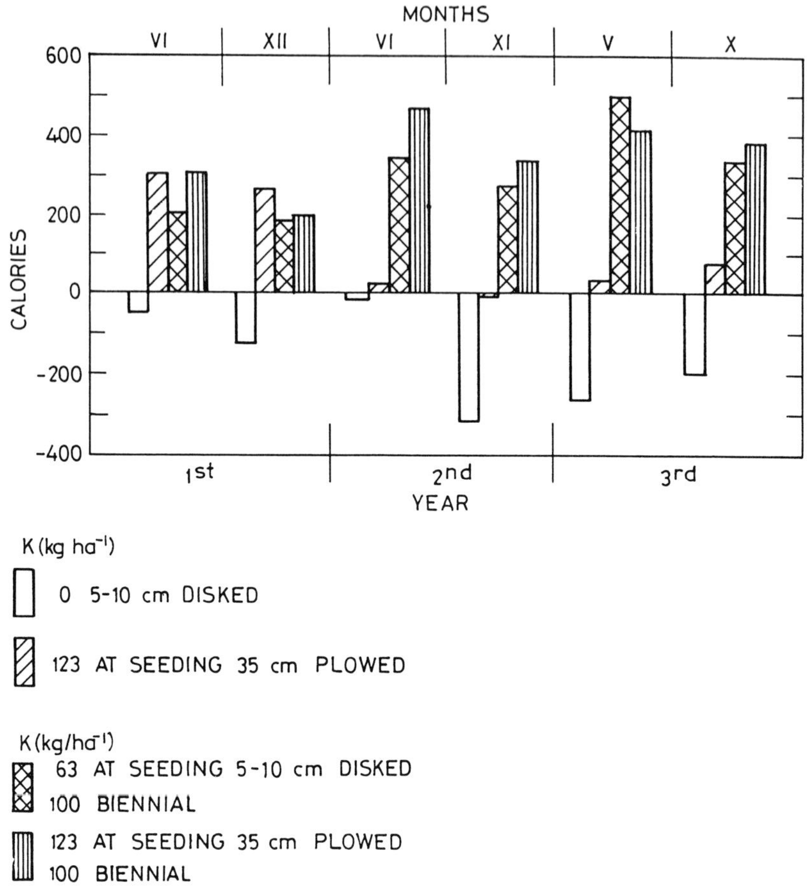

Fig. 5.6. Effect of K fertilizer on the content of available soil-K in the top 0–20 cm soil layer in an alluvial–loam soil in Israel. Figures are the difference between current and initial available soil-K. Soil-K availability denotes the concentration of K in relation to that of Ca and Mg in the water extract of soil, expressed as free energy in calories, according to Woodruff (1955). (Adapted from Levin et al., 1969).

which indicate the concentration range of nutrients in response categories (viz., deficient, critical, adequate and high) are available from California (Meyer and Martin, 1978) and from Australia (Smith, 1986). These manuals also provide details on methods of plant sampling and of chemical analyses. Growing conditions just prior to sampling (such as excess or limited moisture, warm or cool temperatures and disease, particularly root diseases) and the cultivar used, may influence nutrient levels. Of importance is the fact that nutrient composition of some elements changes with

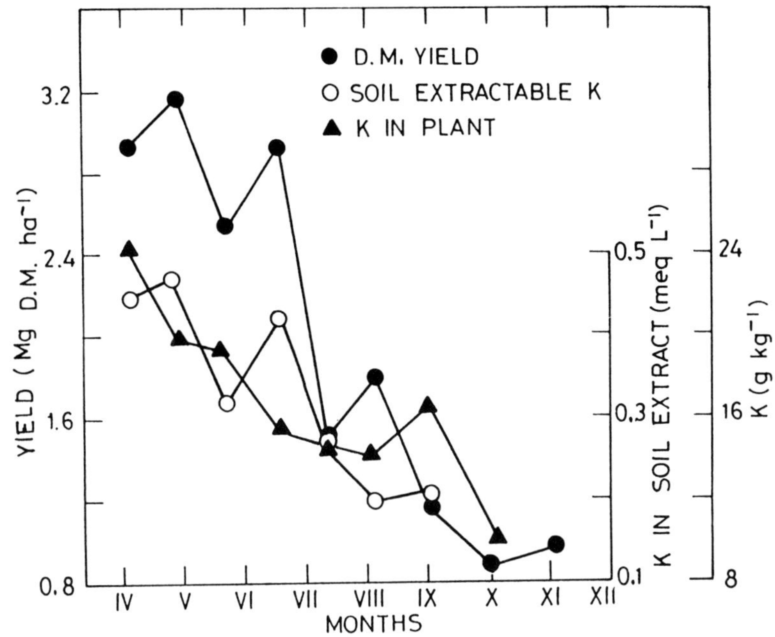

Fig. 5.7. Alfalfa yields in control plots on sandy-loam soil, soil-extractable K in the top 0–20 cm soil layer and the concentration of K in plants. (Adapted from Kafkafi et al., 1977).

TABLE 5.4

Estimate of annual nutrient removal by alfalfa. (Adapted from Martin and Luckhart, 1971)

Element	Removal from 1 Mg of hay (kg ha^{-1})
Nitrogen	225
Potassium	162
Calcium	144
Magnesium	30
Phosphorus	23
Sulfur	22
Iron	1.8
Chlorine	1.4
Manganese	0.9
Boron	0.3
Zinc	0.14
Copper	0.09
Molybdenum	0.009

the plant part sampled and the stage of growth. The concentrations of elements most commonly in short supply and corresponding to different response categories, are shown in Table 5.5. Ranges in nutrient concentration are presented rather than specific values, because the observed relationships between plant yield and concentration of an element in a plant part are not always precise. Responses to fertilization with the nutrient in question can be expected with plant tissue concentrations in the deficient and critical ranges.

TABLE 5.5

Concentrations of nutrients most commonly in short supply in alfalfa of different response categories. Sampling was done at 1/10 bloom stage. (From Meyer and Martin, 1978)

Response category	Total K (middle stems)	PO_4–P (middle stems)	SO_4–S (leaves)	Total Mo (leaves)	Total B (whole tops)
	$g\ kg^{-1}$	$mg\ kg^{-1}$			
Deficient	4–6.5	300–500	<80	<0.3	<15
Critical	6.5–8	501–800	80–150	0.4–0.9	15–20
Adequate	8-15	801–1500	151–500	1.5	21–40
High	15	150	501–1000	5–10[1]	200[2]

[1]Consumption of forage containing more than 10 mg kg^{-1} of Mo may cause 'molybdenosis' in ruminant animals.
[2]Boron concentrations in alfalfa greater than 200 mg kg^{-1} are associated with reduced plant growth and vigor.

Using critical nutrient levels as a basis for fertilizer recommendations requires much experience and good judgment and usually the levels are not the only criteria by which the need for fertilization is assessed. This explains why in the literature the values for critical plant composition data vary considerably since sometimes also safety margins are taken into account. Higher nutrient sufficiency levels in whole irrigated alfalfa plants than those shown in Table 5.5, recommended for maximum growth (depending on growth stage and plant part sampled), are reported for K by Kafkafi et al. (1977) and Smith and Dobrenz (1982); for P by Smith and Dobrenz (1982) and Smith (1986); and for S by Rehm (1987).

The concentration of nutrients in alfalfa leaves is higher than in stems except for K, where the reverse is true. Nutrient concentrations generally decrease as plants develop, and more rapidly in stems than in leaves (Kilcher and Heinrichs, 1974; Rominger et al., 1975).

Yield increases in response to improved management practices, such as in irrigation, fertilizer application, or a greater number of harvests, raise the nutrient uptake of all elements. Examples of the yields and of K and P uptake by alfalfa, including those after fertilizer application at various locations, are given in Table 5.6. A recent tissue survey of irrigated alfalfa in Arizona showed that almost half of the fields were below the minimum sufficiency level for P and most of the fields were below it for Mg. Fertili-

TABLE 5.6

Annual yield and P and K uptake response to fertilization of irrigated alfalfa. Numbers in parentheses are average contents of elements in total herbage (g kg^{-1})

Location, soil type and no. harvests per year	Fertilizer application (kg ha^{-1})	DM yield (Mg ha^{-1})	Uptake (kg ha^{-1}) K	Uptake (kg ha^{-1}) P	Reference
California Silt loam 4		14.5	350 (24)		Meyer (1975)
Israel Sandy loam 9	Control	14.5	276 (19)		Kafkafi et al. (1977)
	600 K as KCl	21.6	562 (26)		
Washington Silt loam 4		14.7	384 (26)		James et al. (1975)
Minnesota Sandy loam 3	Control	9.2	156 (17)	24 (2.6)	Sheaffer (1984)
	501 K as KCl	13.7	399 (29)	30 (2.2)	
California Loam 8(?) harvests	Control	12.7		19 (1.5)	Sallee et al. (1959)
	156 P as superphosphate	17.5		46 (2.6)	
Kansas Sandy loam 6	Control	16.9	394 (23)	43 (2.5)	Ball and Ten Eyck (1980)
	59 P, 149 K	21.9	494 (23)	69 (3.2)	
Colorado Loam, Ravola 4	Control	12.7	253 (20)		Havlin et al. (1984)
	140 K as KCl	13.2	294 (22)		
Idem	Control	12.9		29 (2.2)	Idem
	50 P as superphosphate	13.0		30 (2.3)	

zation with K usually did not result in increased alfalfa hay yields. Significant amounts of N were applied by growers even though tests showed no benefits from this application (Dennis, 1978; Smith and Dobrenz, 1982). In the absence of a yield increase after heavy NPK fertilization, the increase of elemental concentrations in high-yielding alfalfa in southern Peru was rather small, except for P (Versteeg, 1985).

Phosphorus was the most important nutrient required by irrigated alfalfa on an alkaline brown-clay soil in New South Wales (Muldoon, 1986). Omitting N, K, S, or Zn from the 'complete' fertilizer had no effect on DM yields.

Well-inoculated roots of alfalfa fix large amounts of atmospheric N by symbiotic N_2 fixation, usually the principal source of N uptake by alfalfa plants. Estimates of N_2 fixation in alfalfa vary considerably (Vance et al., 1988). The reason for the substantial range in measured values is that only few field measurements of N_2 fixation have been made. N_2 fixation varies since it is affected by many factors, including the bacterial strain, plant genotype interactions (Vance et al., 1988), crop management factors such as soil N (Munns, 1977), soil pH, and K (Collins and Duke, 1981) and P levels. Preliminary data from greenhouse experiments in which plants of nondormant alfalfa were selected for increased N_2 fixation, showed significant increases in yield and quality (Phillips et al., 1981).

5.6 HARVEST MANAGEMENT

Results from many experiments have shown that the number of harvests and the length of intervals between harvests affect total seasonal yield, the quality of the harvested material, and stand persistence. Complementary effects of the frequency of cutting express themselves in plant vigor, rate of recovery after harvest, weediness, and resistance to diseases. Cutting schedules have been based on fixed time intervals, stage of growth, or crown shoot development — offering the producer a variety of harvest options to choose from, depending on production goals and other farm operations.

5.6.1 Regrowth after Cutting

Normally the main source of energy for the initial regrowth after harvest (i.e., production of new stems, leaves and fibrous roots) is total nonstructural carbohydrates (TNC) stored in the root system (including crowns) during the preceding growth cycle (Hodgkinson, 1969; Smith and Silva, 1969; Rains et al., 1981; Denison and Loomis, 1989). This cyclic pattern of utilization and storage of root reserves during the intervals between cuttings, a process that is still not fully understood, is recognized as a key factor in sustained forage production and stand persistence. Graber et al. (1927) were the first to demonstrate that the depletion of root reserves following harvest lasts until the products of photosynthesis manufactured in the new foliage are sufficient to meet the energy requirements for growth.

This depletion of reserves may take one to three weeks, depending on the cultivar, season, and rate of photosynthesis. Full foliage cover on the ground usually marks the onset of the increase in root TNC levels until the next harvest. Frequent harvests of immature alfalfa (vegetative to bud), which prevents the vegetative regrowth from developing enough to restore reserves, can result in reduced root TNC content (Robison and Massengale, 1968) and delayed regrowth (Tabori et al., 1984; Table 5.7). Short intervals between harvests improve the nutritive value of the herbage, but barely affect the total number of cuttings during the season, because of the slower regrowth after harvesting compared with longer intervals between harvests.

TABLE 5.7

Average number of days to reach bud stage, 10% or 60% bloom when irrigated alfalfa was harvested at these growth stages during summer (April–October) in Israel. (Adapted from Tabori et al., 1984)

Stage at harvest	Number of days to reach		
	Bud stage	Flowering	
		10%	60%
Bud stage	20		
10% bloom	17	20	
60% bloom	15	18	20

The reported absolute TNC levels as well as the range of fluctuations of these levels in the roots, are usually smaller in nondormant than in dormant cultivars (Robison and Massengale, 1968; Cohen et al., 1972; R.S. Loomis, unpublished results). This difference in TNC levels between various categories of cultivars is apparently due to the usually more frequent harvests of alfalfa in regions where nondormant cultivars are adapted, because of the warmer climate there, compared with the regions where dormant cultivars are grown.

therefore depends on the allowable soil moisture depletion and differs between soil type and time of the year. If two-thirds of the root mass is concentrated in the top 45 cm of soil, irrigation should supply adequate amounts of water to the soil depth with the greatest concentration of roots (Guitjens, 1990).

Manipulating irrigation schedules when water supplies are limited is feasible because of the outstanding drought resistance of the mature alfalfa plant. Drought-imposed alfalfa quickly recovers after irrigation is restored. Elimination of irrigations should start with the last cutting of the season, then the next-to-last cutting, and so forth.

Since alfalfa is moderately sensitive to salinity, controlling salt effects is important for irrigation waters with salinities exceeding 1 dS m^{-1}. Waste water can be used as an alternative source of fertilizer and soil amendments. The use of low quality water, however, requires practices to control environmental impacts.

Alfalfa should be carefully managed to maintain digestibility. Major preharvest factors affecting forage quality are maturity stage and leaf-to-stem ratio. The concentrations of digestible energy and protein decrease with advancing maturity. This decline in digestibility is faster in the hot summer months than during cooler weather. Harvesting alfalfa in the vegetative to flower-bud stage produces high quality forage, but promotes weed growth and shortens stand life. Whether the grower has to target for maximum quality or quantity of yield depends on the expected financial returns in the light of production costs, which often vary greatly with region, climate, farm, livestock needs and the economic infrastructure.

Field wilting and curing are major sources of harvest losses; storage conditions can also be an additional source of serious nutrient deterioration. Nutrient losses are minimal when animals are fed green chopped alfalfa.

REFERENCES

Abdul-Jabbar, A.S., Sammis, T.W. and Lugg, D.G. 1984. Effect of moisture level on the root pattern of alfalfa. Irrig. Sci. 3: 197–207.

Anon. 1971. Establishing alfalfa stands including methods, rates and dates of seeding. Proc. 1st Calif. Alfalfa Prod. Symp. (Fresno, CA), pp. 14–15.

Anon. 1985. Integrated Pest Management for Alfalfa Hay. Univ. Calif. Publ. 33/2, 96 pp.

Assadian, N.W. and Miyamoto, S. 1987. Salt effects on alfalfa seedling emergence. Agron. J. 79: 710–714.

Baldocchi, D.D., Verma, S.B. and Rosenberg, N.J. 1981. Seasonal and diurnal variation in the CO_2 flux and CO_2 water flux ratio of alfalfa. Agric. Meteorol. 23: 231–244.

Ball, J.A. and Ten Eyck, G. 1980. Top management of irrigated alfalfa produces top yields. Better Crops Plant Food 64: 16–19.

Ball, J.A., Ten Eyck, G. and Nuttleman, R.F. 1980. Irrigated alfalfa research. Kansas State Univ. Agric. Facts, AF-51.

Bath, D.L. and Marble, V.L. 1989. Testing alfalfa for its feeding value. Coop. Ext. Univ. Calif., Div. Agric. Natur. Resour., Leafl. 21457.

Bauder, J.W. and Bauer, A. 1978. Irrigated alfalfa forage productions on a well-drained sandy loam at Oakes, North Dakota, 1974–1977. North Dak. Farm Res. 36: 19–24.

Bauder, J.W., Bauer, A., Ramirez, J.M. and Cassel, D.K. 1978. Alfalfa water use and production on dryland and irrigated sandy loam. Agron. J. 70: 95–99.

Bernstein, L. and Francois, L.E. 1973. Leaching requirement studies: Sensitivity of alfalfa to salinity and drainage waters. Soil Sci. Soc. Am. Proc. 37: 931–943.

Bolger, T.P. and Matches, A.G. 1990. Water-use efficiency and yield of sanfoin and alfalfa. Crop Sci. 30: 143–148.

Bornstein, J., Benoit, G.R., Scott, F.R., Hepler, P.R. and Hedstrom, W.E. 1984. Alfalfa growth and soil oxygen diffusion as influenced by depth to water table. Soil Sci. Soc. Am. J. 48: 1165–1169.

Bresler, E., McNeal, B.L. and Carter, D.L. 1982. Saline and Sodiac Soils. Adv. Ser. Agric. Sci. 10. Springer Verlag, Berlin. pp. 236.

Briggs, L.J. and Shantz, H.L. 1913. The water requirements of plants. I. Investigations in the Great Plains in 1910 and 1911. USDA Bur. Plant Ind. Bull. 282. pp. 44.

Brown, L.G., Hoveland, C.S. and Karnok, K.J. 1990. Harvest management effects on alfalfa yield and root carbohydrates in three Georgia environments. Agron. J. 82: 267–273.

Brown, P.W. and Tanner, C.B. 1983. Alfalfa stem and leaf growth during water stress. Agron. J. 75: 799–804.

Bula, R.J. and Massengale, M.A . 1972. Environmental physiology. In: C.H. Hanson (Ed.) Alfalfa Science and Technology. Agron. Monogr. 15: 167–184.

Buxton, D.R. 1988. Quality potential of alfalfa. Proc. 18th Nation. Alfalfa Symp. (St. Joseph, MI.)

Buxton, D.R. and Hornstein, J.S. 1986. Cell-wall concentration and components in stratified canopies of alfalfa, birdsfoot trefoil, and red clover. Crop Sci. 26: 180–184.

Cameron, D.G. 1973. Lucerne in wet soils — the effect of stage of regrowth, cultivar, air temperature and root temperature. Aust. J. Agric. Res. 24: 851–861.

Chatterton, N.J., Akao, S., Carlson, G.E. and Hungerford, W.E. 1977. Physiological components of yield and tolerance to frequent harvests in alfalfa. Crop Sci. 17: 918–923.

Cohen, Y., Bielorai, H. and Dovrat, A. 1972. Effect of timing of irrigation on total carbohydrate level in roots and on seed yield of alfalfa (*Medicago sativa* L.). Crop Sci. 12: 634–636.

Collins, M. 1983. Changes in composition of alfalfa, red clover and birdsfoot trefoil during autumn. Agron. J. 75: 287–291.

Collins, M. and Duke, S.H. 1981. Influence of potassium — fertilization rate and form on photosynthesis and N_2 fixation of alfalfa. Crop Sci. 21: 481–488.

Conrad, H.R. and Klopfenstein, T.J. 1988. Role in livestock feeding — green chop silage hay and dehy. In: A.A. Hanson, D.K. Barnes and R.R. Hill (Eds.) Alfalfa and Alfalfa

Improvement. ASA, CSSA and SSSA Monogr. 29: 539–550.
Dawson, J.H. and Rincker, C.M. 1982. Weeds in new seedings of alfalfa for seed production: Competition and control. Weed Sci. 30: 20–25.
Day, A.D. Swingle, R.S., Tucker, and Cluff, C.B. 1982. Alfalfa hay grown with municipal waste water and pump water. J. Environ. Qual. 11: 23–-4.
Delaney, R.H., Dobrenz, A.K. and Poole, H.T. 1974. Seasonal variation in photosynthesis, respiration and growth components of non dormant alfalfa (*Medicago sativa* L.). Crop Sci. 14: 58–61.
Denison, R.F. and Loomis, R.S. 1989. An integrative physiological model of alfalfa growth and development. Univ. Calif., Div. Agric. Natur. Resourc. Publ. 1926. 73 pp.
Dennis, R.E. 1978. Key factors for high alfalfa yields in Arizona. 18th Calif. Alfalfa Symp. (Davis, CA). pp. 34–37.
Div. of Agric. and Natural Resources. 1985. Integrated Pest Management for Alfalfa Hay. Univ. of Calif., Publ. 3312. 66 pp.
Donovan, T.J. and Meek, B.S. 1983. Alfalfa responses to irrigation treatment and environment. Agron. J. 75: 461–464.
Doorenbos, J. and Kassam, A.H. 1979. Yield response to water. FAO Irrigation Drainage Paper 33, FAO Rome. 193 pp.
Doorenbos, J. and Pruitt, W.O. 1977. Guidelines for predicting crop water requirements. FAO Irrigation Drainage Paper 24, FAO Rome. 144 pp.
Dovrat, A. 1962. Potassium deficiency as a result of increased nitrogen fertilization in irrigated pastures. I.P.I. Potassium Symposium (Athens, Greece), Vol. 7, pp. 375–385.
Droushiotis, D.N. 1980. The effects of the stage of growth and height on the yield of irrigated lucerne. Nicosia Agric. Inst. Tech. Bull. 36.
Eck, H.V., Martinez, T. and Wilson, G.C. 1977. Alfalfa production on a profile-modified slowly permeable soil. J. Soil Sci. Soc. Am. 41: 1181–1186.
Elmore, C. 1987. Growing alfalfa on heavy clay soil with beds. Proc. 17th California/Arizona Alfalfa Symp. (El Centro, CA) pp. 69–70.
Evans, D.W. and Peaden, R.N. 1984. Seasonal forage growth rate and solar energy conversion of irrigated Vernal alfalfa. Crop Sci. 24: 981–984.
Feltner, K.C. and Massengale, M.A. 1965. Influence of temperature and harvest management on growth, level of carbohydrates in the roots and survival of alfalfa (*Medicago sativa* L.). Crop Sci. 5: 585–588.
Fick, G.W., Holt, D.A. and Lugg, D.G. 1988. Environmental physiology and crop growth. In: A.A. Hanson, D.K. Barnes and R.R. Hill (Eds.) Alfalfa and Alfalfa Improvement. ASA, CSSA and SSSC Monogr. 29: 163–194.
Finkel, H.J. 1982. Irrigation of alfalfa. In: H.J. Finkel (Ed.) Handbook of Irrigation Technology. Vol. II. pp. 191–197. CRC Press Inc., Boca Raton, FL.
Fixen, P.E. and Ludwick, A.E. 1983. Phosphorus and potassium fertilization of irrigated alfalfa on calcareous soils. I. Soil test maintenance requirements. Soil Sci. Soc. Am. J. 47: 107–112.
Francois, L.E. 1981. Alfalfa management under saline conditions with zero leaching. Agron. J. 73: 1042–1048.
Frate, C.A., Roberts, B.A. and Marble, V.L. 1991. Imposed drought stress has no long-term effect on established alfalfa. Calif. Agric. 45 (3): 33–36.

Free, J.B. 1970. Insect Pollination of Crops, Academic Press, London. 544 pp.
Goldman, A. and Bar-Cohen, R. 1975. Efficiency of honey bees as pollinators of alfalfa for seed. Bull Agric. Res. Orgn. Bet Dagan 150 (In Hebrew).
Goldman, A. and Devash, L. 1990. A control system for improving pollination efficiency of alfalfa by honey bees. Israel Agresearch 4: 85–101.
Goldman, A. and Dovrat, A. 1980. Irrigation regime and honey bee activity as related to seed yield in alfalfa. Agron. J. 72: 961–965.
Graber, L.F., Nelson, N.T., Luckel, W.A. and Albert, W.B. 1927. Organic food reserves in relation to the growth of alfalfa and other perennial herbaceous plants. Wisc. Agric. Ext. Stn. Bull. 80.
Graham, J.H., Froheiser, F.I., Stuteville, D.L. and Erwin, D.C. 1979. A Compendium of Alfalfa Diseases. Am. Phytopathol. Soc., St. Paul, MN, 65 pp.
Gregory, E.J., Ferris, C. and Marble, V.L. 1965. Alfalfa seed production in Fresno County. Calif. Agric. Exp. Stn. Fresno County Circ.
Grimes, D.W., Sheesley, W.R. and Wiley, P.L. 1978. Alfalfa root development and shoot regrowth in compact soil of wheel traffic patterns. Agron. J. 70: 955–958.
Grimes, D.W., Wiley, P.L. and Sheesley, W.R. 1992. Alfalfa yield and plant relations with variable irrigation. Crop Sci. 32: 138–1387.
Guessous, F., Igmoullan, A. and Johnson, W.L. 1986. Nutritive value of alfalfa produced in a Mediterranean climate. J. Anim. Sci. Suppl. 1, p. 308.
Guitjens, J.C. 1982. Models of alfalfa yield and evapotranspiration. J. Irrig. Drain. Div., Proc. Am. Soc. Civil Eng. 108, IR3: 212.
Guitjens, J.C. 1990. Alfalfa. In: B.A. Stewart and D.R. Nielsen (Eds.) Irrigation of Agricultural Crops. ASA, CSSA and SSSA Monogr. 30: 537–568.
Hagemann, R.W. and Marble, V.L. 1983. Variety response to cutting schedules in Imperial Valley. Proc. 13th Calif. Alfalfa Symp. (Holtville, Univ. Calif. Coop. Ext.), pp. 6–15.
Hagemann, R.W., Willardson, L.S., Marsh, A.W. and Ehlig, C.F. 1975. Irrigating for maximum alfalfa seed yield. Calif. Agr. 29 (11): 14–15.
Hagin, J. and Dovrat, A. 1963. Note on methods for determination of available soil potassium. Emp. J. Exp. Agric. 31: 186–188.
Hall, M.H. and Eckert, J.W. 1991. Decreased alfalfa growth after water-deficit stress is removed. Agron. Abstr., ASA, CSSA and SSSA (Denver, CO). p. 145.
Hanson, A.A., Barnes, D.K. and Hill, R.R. (Eds.) 1988. Alfalfa and Alfalfa Improvement. ASA, CSSA and SSSA Monogr. 29. 1084 pp.
Hanson, B. 1991. Alfalfa irrigation management. Proc. 21st Calif. Alfalfa Symp. (Davis, CA.) pp. 70–80.
Hattendorf, M.J., Evans, D.W. and Peaden, R.N. 1990. Canopy temperature and stomatal conductance of water-stressed dormant and nondormant alfalfa types. Agron. J. 82: 873–877.
Havlin, J.L., Westfall, D.G. and Golus, H.M. 1984. Six years of phosphorus and potassium fertilization in irrigated alfalfa on calcareous soils. Soil Sci. Soc. Am. J. 48: 331–336.
Heichel, G.H. 1983. Alfalfa. pp. 127–155. In: I.D. Teare and M.M. Peet (Eds.) Crop–Water Relations, John Wiley & Sons, New York, N.Y.
Heichel, G.H., Delaney, R.H. and Cralle, H.T. 1988. Carbon assimilation, partitioning

and utilization. In: A.A. Hanson, D.K. Barnes and R.R. Hill (Eds.) Alfalfa and Alfalfa Improvement. ASA, CSSA and SSSA Monogr. 29: 195–228.

Henderson, D.W. and Yamada, H. 1979. Irrigation management for alfalfa seed production. Univ. Calif. West Side Field Stn. Annu. Rep.

Henderson, D.W. Yamada, H., and Hoover, R.M. 1975. Irrigating alfalfa for seed. Proc. Calif. Alfalfa Seed Prod. Symp. (Fresno, CA.), pp. 26–29.

Hesterman, O.B. and Teuber, L.R. 1981. Effect of photoperiod and irradiance on fall dormancy of alfalfa. Agron. Abstr., ASA, CSSA and SSSA, Madison, WI. p. 87.

Hodgkinson, K.C. 1969. The utilization of root organic compounds during the regeneration of lucerne. Austr. J. Biol. Sci. 22: 1113–1123.

Hodgkinson, K.C. 1973. Establishment and growth of shoots following low and and high cutting of lucerne in relation to the pattern of nutrient uptake. Austr. J. Agric. Res. 24: 497–510.

James, D.W., Weaver, W.H., Roberts, S. and Hunter, A.H. 1975. Potassium in an arid loessial soil: Changes in availability as related to cropping and fertilization. Soil Sci. Soc. Am. Proc. 39: 1111–1115.

Jensen, M.E. (Ed) 1980. Design and Operation of Farm Irrigation Systems. ASAE Monogr. 3: 551–572.

Jones, L.G. and Pomeroy, C.R. 1962. Effect of fertilizer, row spacing and clipping on alfalfa seed production. Calif. Agric. 16 (2): 8–10.

Judd, P. and Radcliffe, J.C. 1970. The influence of cutting frequency on the yield, composition and persistence of irrigated lucerne. Aust. J. Exp. Agric. Anim. Husb. 10: 48–52.

Kaddah, M.J. 1977. Soil profile modification to enhance alfalfa production. Proc. 7th Calif. Alfalfa Symp. (Fresno, CA), pp. 22–24.

Kafkafi, U., Gilat, R., Yoles, D. and Noy, Y. 1977. Studies on fertilization of field-grown irrigated alfalfa. I. Effect of potassium source and time of application. Plant and Soil 46: 165–173.

Kalu, B.A. and Fick, G.W. 1981. Quantifying morphological development of alfalfa for studies of herbage quality. Crop Sci. 21: 267–271.

Kalu, B.A. and Fick, G.W. 1983. Morphological stage of development of alfalfa as a predictor of herbage quality. Crop Sci. 23: 1167–1172.

Kilcher, M.R. and Heinrichs, D,H. 1974. Contribution of stems and leaves to the yield and nutrient level of irrigated alfalfa of different stages of development. Can. J. Plant Sci. 54: 739–742.

Kipnis, T., Vaisman, I. and Granoth, I. 1989. Drought stress and alfalfa production in a Mediterranean environment. Irrig. Sci. 10: 113–125.

Lanini, W.T., Orloff, S.B., Vargas, R.N., Orr, J.P., Marble, V.L. and Grattan, S.R. 1991. Oat companion crop seeding rate on alfalfa establishment, yield and weed control. Agron. J. 83: 330–333.

Lanyon, L.E. and Griffith, W.K. 1988. Nutrition and fertilizer use. In: A.A. Hanson, D.K. Barnes and R.R. Hill (Eds.). Alfalfa and Alfalfa Improvement. AASA, CSSA and SSSA Monogr. 29: 333–372.

Leach, G.J. 1969. Plant numbers, shoot size and yield of regrowth in three lucerne cultivars. Aust. J. Agric. Res. 20: 425–434.

Leach, G.J. 1979. Regrowth characteristics of lucerne under different systems of grazing management. Aust. J. Agric. Res. 30: 445–465.

Leath, K.T., Erwin, D.C. and Griffin, G.D. 1988. Diseases and nematodes. In: A.A. Hanson, D.K. Barnes and R.R. Hill (Eds.) Alfalfa and Alfalfa Improvement. ASA, CSSA and SSSA. Monogr. 29: 621–670.

Leavitt, J.R.C., Dobrenz, A.K. and Stone, J.E. 1979. Physiological and morphological characteristics of large and small leaflet alfalfa genotypes. Agron. J. 71: 529–532.

Lehman, W.F. 1979. Alfalfa production in the low desert valley areas of California. Univ. Calif. Div. Agr. Sci. Leafl. 21097.

Lehman, W.F., Marble, V.L., Erwin, D.C. and Nielson, M.W. 1987. Nondormant alfalfa, past, present and future. Proc. 17th Calif./Ariz. Alfalfa Symp. (El Centro, CA), p. 29

Levin, I., Dovrat, A. and Mozes, G. 1969. Yield responses of irrigated alfalfa (*Medicago sativa* L.) to available soil P and K levels in different fertilizer treatments in the Hula Valley of Israel. Israel J. Agric. Res. 19: 25–30.

Lodge, G.M. 1986. Yield and persistence of irrigated lucernes cut at different frequencies at Tamworth, New South Wales. Aust. J. Exp. Agric. 26: 165–172.

Loomis, R.S. and Wallinga, J. 1991. Alfalfa: Efficient or ineffcient use of water. Proc. 21st Calif. Alfalfa Symp. (Sacramento, CA), pp. 63–69.

Lowe, K.F., Gramshaw, D., Bowdler, T.M. and Ludke, D.H. 1985. Performance of North American and Australian lucernes in the Queensland subtropics. 2. Yield and plant survival in irrigated stands. Aust. J. Exp. Agr. 25: 82–90.

Ludlow, M.M. 1980. Adaptive significance of stomatal response to water stress. p. 123. In: N.C. Turner and P.J. Kramer (Eds.) Adaptation of Plants to Water and High Temperature Stress. John Wiley & Sons, New York, N.Y.

Maas, E.V. 1987. Salt tolerance of plants. In: B.R. Christie (Ed.). pp. 57–75. Handbk. Plant Science Agriculture II, CRC Press, Boca Raton, FL.

Manglitz, G.R. and Ratcliffe, R.H. 1988. Insects and mites. In: A.A. Hanson, D.K. Barnes and R.R. Hill (Eds.), Alfalfa and Alfalfa Improvement. ASA, CSSA and SSSA. Monogr. 29: 671–704.

Marble, V.L. 1970. Producing alfalfa seed in California. Univ. Calif. Davis, Div. Agric. Sci. Leafl. 2382.

Marble, V.L. 1972. Optimizing alfalfa production in California. Proc. 2nd Calif. Alfalfa Symp. (Fresno, CA), pp. 41–55.

Marble, V.L. 1974. How cutting schedules and varieties affect yield, quality and stand life. Proc. 4th Calif. Alfalfa Symp. (Fresno, CA), pp. 47–57.

Marble, V.L. 1980. Effect of harvest frequencies and variety on yield quality and stand life. Proc. 10th Calif. Alfalfa Symp. (Visalia, CA), pp. 22–38.

Marble, V.L. 1984. A new system for determining alfalfa hay quality: Acid detergent fiber. Proc. 14th Calif. Alfalfa Symp. pp. 22–42.

Marble, V.L. 1986. Relative advantages of different dormancies of alfalfa grown in central and northern California. Proc. 16th Calif. Alfalfa Symp. (Sacramento, CA), pp. 6–35.

Marble, V. L. 1989. Fodders for the Near East: Alfalfa. FAO Plant Prod. Protec. Paper 97/1. pp. 207.

Marble, V.L., Peterson, G. and Shoner, C.A. 1988. Effect of fall/winter cutting on alfalfa (*Medicago sativa* L.) yield, pest management and stand life. Proc. 18th Calif. Alfalfa Symp. (Modesto, CA), pp. 45–58.

Marble, V.L., Baghott, K.G., Benton, R.W., Smith, P.D. and Gripp, R.H. 1985a. Producing quality alfalfa in California's mountain valleys. Calif. Agric. 39: 27–30.

Marble, V.L., Wilton, A.C. and Cook, A.A. 1985b. Performance of nine mid-east alfalfa ecotypes and seventeen introduced cultivars in Saudi Arabia. Proc. 15th Int. Grassland Congr. (Kyoto, Japan), pp. 129–131.

Marten, G.C. 1970. Temperature as a determinant of quality of alfalfa harvested by bloom or age criteria. Proc. 11th Int. Grassland Congr. (Surfers Paradise, Australia), pp. 506–509.

Marten, G.C., Buxton, D.R. and Barnes, R.F. 1988. Feeding value (Forage quality). In: A.A. Hanson, D.K. Barnes, and R.R. Hill (Eds.), Alfalfa and Alfalfa Improvement. ASA, CSSA and SSSA Monogr. 29: 463–491.

Martin, W.E. and Luckhart, R.L. 1971. How to predict fertilizer needs of alfalfa. Calif. Alfalfa Prod. Symp. (Fresno, CA), pp. 16–18.

Martinez, R.A., Quiroga, M., Thomas, N. and Byerly, K.F. 1983. Alternative forage-production patterns for dairy cattle under conditions of limited water supply in Northern Mexico. Proc. 16th Int. Grassland Congr. (Lexington, KY). pp. 804–806.

Massengale, M.A. 1974. Management of alfalfa for increased forage production. Proc. California/Arizona low desert Alfalfa Symp. (El Centro, CA). pp. 69–71.

McKenzie, J.S., Paquin, R. and Duke, S.H. 1988. Cold and heat tolerance. In: A.A. Hansen, D.K. Barnes, and R.R. Hill (Eds.), Alfalfa and Alfalfa Improvement. ASA, CSSA and SSSA Monogr. 29: 259–302.

McWilliams, J.R. 1980. Adaptation of plants to water and high temperature stress, pp. 444–447. In: N.C. Turner and P.J. Kramer (Eds.), Adaptation of Plants to Water and High Temperature Stress, John Wiley & Sons, New York.

Meek, B.D. 1978. Effect of deep tillage on alfalfa yield. pp. Proc. 8th Calif. Alfalfa. Symp. (Davis, CA). pp. 85–86.

Meek, B.D., Rechel, E.A., Carter, L.M. and DeTar, W.R. 1988. Soil composition and its effect on alfalfa in zone production systems. Soil Sci. Soc. Am. J. 52: 233–236.

Metochis, C. 1980. Irrigation of lucerne under semi-arid conditions in Cyprus. Irrig. Sci. 1: 247–252.

Metochis, C. and Orphanos, P.I. 1981. Alfalfa yield and water use when forced into dormancy by withholding water during the summer. Agron. J. 73: 1048–1050.

Meyer, J.L. 1971. Soil management production as they relate to soil types and alfalfa growth. Proc. 1st Calif. Alfalfa Symp. (Fresno, CA), pp. 1–4.

Meyer, R.D. 1975. Detecting nutritional needs of alfalfa. Proc. 5th Calif. Alfalfa Symp. (Berkeley, CA), pp. 79–81.

Meyer, R.D. and Martin, E.W. 1978. Plant analysis as a guide for fertilization of alfalfa. pp. 27–29. In: H.M. Reisenauer (Ed.) Soil and Plant-Tissue Testing in California. Div. Agr. Sci. Univ. Calif. Bull. 1879.

Meyerhoff, P.A. 1984. Growth of established alfalfa: Relationship to changes in fibrous root mass and response to nitrogen fertilization. M.Sc. thesis, Univ. of California. Davis, CA. 122 pp.

Mitchell, A.R. and Swain, R.V. 1987. Controlled traffic on alfalfa in the Imperial Valley — a whole new ball game. Proc. 17th California/Arizona Alfalfa Symp. (El Centro, CA), pp. 71–76.

Mokeyera, E.V. 1940. Biological–anatomical studies of alfalfa (*Medicago sativa*). (Eng-

lish translation). State Agric. Publ. House Ltd. Tashkent, USSR.
Muldoon, D.K. 1986. Nutrition of irrigated crops on an alkaline brown clay soil at Trangie, New South Wales. I. Lucerne. Aust. J. Exp. Agric. 26: 245–249.
Munns, D.N. 1977. Mineral nutrition and the legume symbiosis. pp. 353–391. In: R.W.F. Hardy and A.H. Gibson (Eds.). A Treatise on Dinitrogen Fixation IV: Agronomy and Ecology. John Wiley & Sons, New York, NY.
Norris, R.F. 1980. Season-long yield losses from weeds in seeding stands. Proc. 10th Calif. Alfalfa Symp. (Visalia, CA), pp. 68–74.
Olmstead, A.L. and Wooten, D.B. 1987. Bee pollination and productivity growth: The case of alfalfa. Am. J. Agric. Econ. 69: 56–63.
Olsen, S.R. 1954. Estimation of available phosphorus in soils by extraction with $NaHCO_3$. U.S. Dep. Agric. Circ. 939.
Onstad, D.W. and Fick, G.W. 1983. Predicting crude protein, *in vitro* true digestibility, and leaf proportion in alfalfa herbage. Crop Sci. 23: 961–964.
Oron, G., De Malach, Y. and Hoffman, Z. 1989. Sub-surface trickle irrigation of alfalfa with treated waste water. Hassadeh 69: 2087–2092 (in Hebrew).
Pauli, A.W., Lechtenberg, V.L. and Wedin, W.F. 1988. Equipment for harvesting, transporting, storing and feeding. In: A.A. Hanson, D.K. Barnes and R.R. Hill (Eds.). Alfalfa and Alfafa Improvement. ASA, CSSA and SSSA. Soc. Agron. Monogr. 29: 567–594.
Perry, L.J. and Larson, K.L. 1974. Influence of drought on tillering and internode number and length in alfalfa. Crop Sci. 14: 693–696.
Peters, E.J. and Linscott, D.L. 1988. Weeds and weed control. In: A.A. Hanson, D.K. Barnes and R.R. Hill (Eds.). Alfalfa and Alfafa Improvement. ASA, CSSA and SSSA Monogr. 29: 705–735.
Pfeiffer, T., Schrader, L.E. and Bingham, E.T. 1980. Physiological comparison of isogenic diploid–tetraploid, tetraploid–octoploid alfalfa populations. Crop Sci. 20: 299–303.
Phillips, D.A., Teuber, L.R. and Jue, S.S. 1981. Variation among alfalfa genotypes for reduced nitrogen concentration. Crop Sci. 22: 606–610.
Rains, D.W. et al. 1981. Systems analysis in alfalfa production. Proc. 11th Calif. Alfalfa Symp. (Fresno, CA). pp. 44–48.
Rammah, A.M. and Hamza, A.S. 1980. Cutting schedules and seasonal effects on yield and quality of alfalfa. Ann. Agric. Soc. Moshtohor, Egypt 14: 61–74.
Rechel, E.A., Carter, L.M., and DeTar, W.R. 1987. Alfalfa growth response to a zone-production system. I. Forage production characteristics. Crop Sci. 27: 1029–1034.
Rechel, E.A., DeTar, W.R., Meek, B.D. and Carter, L.M. 1991. Alfalfa (*Medicago sativa* L.) water use efficiency as affected by harvest traffic and soil compaction in a sandy loam soil. Irrig. Sci. 12: 61–65.
Rehm, G.W. 1987. Application of phosphorus and sulphur on irrigated alfalfa. Agron. J. 79: 973–979.
Reisenauer, H.M., Quick, J., Voss, R.E. and Brown, A.L. 1978. Soil test interpretive guides. pp. 38–40. In: H.M. Reisenauer (Ed.). Soil and Plant-Tissue Testing in California. Univ. Calif. Davis Div. Agric. Sci. Bull. 1879.
Retta, A. and Hanks, R.J. 1980. Corn and alfalfa production as influenced by limited irrigation. Irrig. Sci. 1: 135–147.

Rincker, C.M., Marble, V.L., Brown, D.E. and Johansen, C.A. 1988. Seed production practices. In: A.A. Hanson, D.K. Barnes and R.R. Hill (Eds.). Alfalfa and Alfalfa Improvement. ASA, CSSA and SSSA Monogr. 29: 985–1021.

Robinson, F.E., McCoy, O.D., Worker, G.P. and Lehman, W.F. 1968. Sprinkler and surface irrigation of vegetables and field crops in an arid environment. Agron. J. 60: 696–699.

Robison, G.D. and Massengale, M.A. 1968. Effect of harvest management and temperature on forage yield, root carbohydrates, plant density and leaf area relationships in alfalfa (*Medicago sativa* L. cv. Moapa). Crop Sci. 8: 147–151.

Rogers, V.E. 1969. Depression of nitrogen uptake and growth of lucerne at high soil temperatures. CSIRO Div. Plant Ind. Field Stat. Rec. 8: 37–44. Canberra.

Rominger, R.S., Smith, D. and Peterson, L.A. 1975. Yields and elemental composition of alfalfa plant parts at late bud under two fertility levels. Can. J. Plant Sci. 55: 69–75.

Rumbaugh, M.D. 1963. Effects of population density on some components of yield of alfalfa. Crop Sci. 3: 423–424.

Sallee, W.R., Ulrich, A., Martin, W.E. and Krantz, B.A. 1959. High phosphorus for alfalfa. Calif. Agric. 13: 7–8.

Sharratt, B.S., Reicosky, D.C., Idso, S.B. and Baker, D.G. 1983. Relationships between leaf water potential, canopy temperature and evapotranspiration in irrigated and non-irrigated alfalfa. Agron. J. 75: 891–894.

Sheaffer, C.C. 1984. Potash for irrigated alfalfa boosts yield and reduces stand loss. Better Crop Plant Food 68: 8–9.

Sheaffer, C.C., Tanner, C.B. and Kirkham, M.B. 1988. Alfalfa water relations and irrigation. In: A.A. Hanson, D.K. Barnes and R.R. Hill (Eds.). Alfalfa and Alfalfa Improvement. ASA, CSSA and SSSA Monogr. 29: 373–402.

Sheehy, J.E., Fishbeck, K.A. and Phillips, D.A. 1980. Relationships between apparent nitrogen fixation and carbon exchange rate in alfalfa. Crop Sci. 20: 491–495.

Sheesley, B. and Farley, J. 1982. Using harvest schedules to adjust quality of summer-grown alfalfa in the Central Joaquin Valley. 12th Calif. Alfalfa Symp. (Davis, CA) pp. 10–14.

Sheesley, B. and Mueller, S. 1988. Producing alfalfa seed in a water-short year. Coop. Ext., Univ. Calif. Fresno. 4 pp.

Sheesley, W.R. 1982. Timing of first alfalfa harvest to avoid impedance of root growth. Proc. 12th Calif. Alfalfa Symp. pp. 47–49.

Sheesley, W.R. 1983. Interdependence of pest control, pollination and irrigation management in alfalfa seed production. Proc. Calif. Alfalfa Seed Production Symp. (Fresno, CA), pp. 3–8.

Sheesley, W.R. 1985. Alternate block irrigation management for alfalfa seed production. Proc. Calif. Alfalfa Seed Production Symp. (Fresno, CA), p. 15.

Sheesley, W.R. and Atkins, E.L. 1985. Value of in-field honey bee watering sites in alfalfa seed pollination. Proc. Calif. Alfalfa Seed Production Symp. (Fresno, CA), pp. 13–14.

Sheesley, W.R., Grimes, D.W., McClellan, W.S., Summers, C.G. and Marble, V.L. 1982. The influence of wheel traffic on yield and stand longevity in alfalfa. Calif. Agric. 28 (10): 6–8.

Simpson, J.R. and Lipsett, J. 1973. Effects of surface moisture-supply on the subsoil nutritional requirements of lucerne (*Medicago sativa* L.). Aust. J. Agric. Res. 24: 199–209.

Slarke, R.H. and Mason, W.K. 1987. Effect of growth stage at cutting on yield and quality of lucerne cultivars from different dormancy groups in northern Victoria. Aust. J. Exp. Agric. 27: 55–58.

Smeal, D., Kallsen, C.E. and Sammis, T.W. 1991. Alfalfa yield as related to transpiration, growth stage and environment. Irrig. Sci. 12: 79–86.

Smith, D. 1961. Association of fall growth habit and winter survival in alfalfa. Can. J. Plant Sci. 41: 244–251.

Smith, D. and Dobrenz, A.K. 1982. What are nutrient needs of high yield alfalfa? Better Crops Plant Food 68: 31–33.

Smith, D. and Silva, J.P. 1969. Use of carbohydrate and nitrogen root reserves in the regrowth of alfalfa from greenhouse experiments under light and dark conditions. Crop Sci. 9: 464–467.

Smith, F.W. 1986. Pasture species. pp. 109–110. In: D.J. Reuter and J.B. Robinson (Eds.). Plant Analysis: An Interpretation Manual. Inkata Press, Melbourne, Australia.

Smith, S.E., Al-Doss, A. and Conta, D.M. 1989. Classification of Middle Eastern alfalfa based on analysis of agronomically important characteristics. Abstr. Annu. Meet. ASA, CSSA and SSSA (Las Vegas, NV), p. 100.

Stegman, E.C., Musick, J.C. and Stewart, J.I. 1980. Irrigation management. In: M.E. Jensen (Ed.) Design and Operation of Irrigation Systems. ASAE Monogr. 3: 763–816.

Stuteville, D.L. and Erwin, D.C (Eds.) 1990. A Compendium of Alfalfa Diseases. Am. Phytopathology. Soc. Press, St Paul, MN 84 pp.

Tabori, C., Sharet, N. and Giladi, J. 1984. Yield and forage quality of alfalfa cut at different stages of development. Hassadeh 64: 1317–1321 (in Hebrew).

Taylor, A.J. 1987. Lucerne seedling growth and first year productivity in south-eastern Australia as influenced by sowing date in autumn and early winter. Aust. J. Exp. Agric. 27: 833–839.

Taylor, A.J. and Marble, V.L. 1986. Lucerne irrigation and soil water use during bloom and seedset on a red-brown earth in south-eastern Australia. Aust. J. Exp. Agric. 26: 577–581.

Taylor, A.J. and Olsson, K.A. 1987. Effect of gypsum and deep ripping on lucerne (*Medicago sativa* L.) yields on a red-brown earth under flood and spray irrigation. Aust. J. Exp. Agric. 27: 841–849.

Tesar, M.B. and Marble, V.L. 1988. Alfalfa establishment. In: A.A. Hanson, D.K. Barnes and R.R. Hill (Eds.). Alfalfa and Alfalfa Improvement. ASA, CSSA and SSSA Monogr. 29: 303–329.

Teuber, L.R. and Brick, M.A. 1988. Morphology and anatomy. In: A.A. Hanson, D.K. Barnes and R.R. Hill (Eds.). ASA, CSSA and SSSA Monogr. 29: 125–162.

Teuber, L.R., Jernstedt, J. and Foord, K. 1987. Alfalfa growth and development. 17th Calif. Alfalfa Symp. (El Centro, CA), pp. 63–68.

Thompson, T.E. and Fick, G.W. 1981. Growth response of alfalfa to duration of soil flooding and to temperature. Agron. J. 73: 329–332.

Valdivia, H. and Zipori, I. 1982. Determinacion del coeficiente de transpiracion de varios cultivos. In: I. Zipori, M. Versteeg and H. Valdivia (Eds.). Factores que influen en la produccion de cultivos alimenticos y forrageros en areas deserticas. FAPROCAF-INIPA, Arequipa, Peru. pp. 193–225.

Van Bavel, C.H.M. 1967. Changes in canopy resistance to water loss from alfalfa induced by water depletion. Agric. Meteorol. 4: 165–176.

Vance, C.P., Heichel, G.H. and Phillips, D.A. 1988. Nodulation and symbiotic dinitrogen fixation. In: A.A. Hanson, D.K. Barnes and R.R. Hill (Eds.). Alfalfa and Alfalfa Improvement. ASA, CSSA and SSSA Monogr. 29: 229–258.

Veronesi, F., Mariani, A., Falcinelli, M. and Arcioni, S. 1986. Selection for tolerance to frequent cutting regimes in alfalfa. Crop Sci. 26: 58–61.

Versteeg, M.N. 1985. Factors influencing the productivity of irrigated crops in southern Peru in relation to prediction by simulation models. Ph.D thesis, Agric. Univ., Wageningen, the Netherlands, pp. 181.

Volenec, J.J., Cherney, J. and Johnson, K.D. 1987. Yield components, plant morphology, and forage quality of alfalfa as influenced by plant population. Crop Sci. 27: 321–326.

Vough, L.R. and Marten, G.C. 1971. Influence of soil moisture and ambient temperature on yield and quality of alfalfa forage. Agron. J. 63: 40–42.

Waldman, M., Dovrat, A. and Shohat, J. 1975. New alfalfa varieties — How good are they? Hassadeh 45: 523–527 (in Hebrew).

Whitfield, D.M., Wright, G.C., Gyles, O.A. and Taylor, A.J. 1986. Growth of lucerne (*Medicago sativa* L.) in response to frequency of irrigation and gypsum application on a heavy clay soil. Irrig. Sci. 7: 37–52.

Woodruff, C.M. 1955. The energies of replacement of calcium by potassium in soils. Proc. Soil Sci. Soc. Am. 19: 167–171.

Woodruff, C.M. 1967. Crop response to lime in the midwestern United States. In: R.W. Pearson and F. Adams (Eds.). Soil Acidity and Liming. SSSA Monogr. 12: 207–231.

Worker, G.F. and Marble, V.L. 1987. Effects of fall and winter harvests on alfalfa yield and quality. 17th California/Arizona Alfalfa Symp. (El Centro, CA), pp. 41–44.

Wright, G.C., Whitfield, D.M., Gyles, O.A. and Taylor, A.J. 1986. Effects of stage of growth, irrigation frequency and gypsum treatment and CO_2 exchange on lucerne (*Medicago sativa* L.) growth on heavy clay soil. Irrig. Sci. 7: 169.

Zipori, I. and Valdivia, L.H. 1982. La repuesta del cultivo de alfalfa (*Medicago sativa* L.) a la aplicacion de agua en zonas aridas. In: I. Zipori, M.N. Versteeg and L.H. Valdivia (Eds.). Factores que influen en la produccion de cultivos alimenticos y forrageros en areas deserticas. FAPROCAF-INIPA, Arequipa, Peru, pp. 243–280.

Chapter 6

IRRIGATED PASTURE

6.1 INTRODUCTION

Justification for the allocation of irrigation water to pastures is usually based on:

(i) greater net returns from animal production by grazing than from feeding cut fodders or from growing cash crops;

(ii) the use of less productive land;

(iii) the use of low quality water;

(iv) providing an effective ground cover for easily-erodible soil or steep slopes;

(v) prolonging the effective growing period during dry seasons; and/or

(vi) supplementing dry annual range.

Documentation of the use of irrigated pastures is available mainly from Australia and North America. The largest irrigated developments in Australia are in the southeastern region (Murray–Darling Basin), where more than 55% of the national irrigated area (from a 1976/77 survey) is devoted to irrigated pastures extending over more than 1 million ha (Smith et al., 1983; Pritchard et al., 1989). The pastures are categorized as perennial, annual and native producing ca. 12, 5 and 2 Mg DM ha^{-1} $year^{-1}$, respectively. The large commitment of irrigation water to pastures was due initially to the dominance of the grazing industry in the early years. Subsequently, however, it has been reinforced by the incapability of most soils to support crops which require permeable well-drained soil conditions. Most surface-irrigated pastures in Victoria, southern Queensland and New South Wales are utilized for high-value milk production (Radcliffe, 1986).

Pasture irrigation in North America originates from the practise by early settlers of diverting water run-off from melting snows or from rivers and streams to adjacent land, commonly referred to as "wild flooding" (Nichols and Clanton, 1985). The extent of irrigated pastures in the Pacific

Coast states of the USA in 1980 was estimated at ca. 750,000 ha, a decrease of almost one-fourth in area compared with 1970. Most of the pastures in the region are used today for beef production (Marble et al., 1985). The average annual DM production is estimated at ca. 9 Mg ha^{-1}. An estimated one-third of the irrigated perennial pastures in California was at one time used for dairy cattle, but the figure is considerably less now, since increased land values have demanded higher yielding crops. Furthermore, the trend toward hay and silage for feed lot dairying, and the complex pasture management required with large herds, also contributed to a shift away from irrigated pastures.

Most of the underlying principles and practices associated with the use of rain-fed pastures hold true for irrigated pastures, since they:

(i) are sown with perennial, annual grasses and legumes (self-reseeding or sown annually) either alone or in mixture;

(ii) provide year-round or seasonal grazing, or are used as a complementary crop depending on rainfall, availability of irrigation water and the forage program; excess growth may be harvested as hay or silage;

(iii) serve in certain farming systems as a soil fertility-building phase in the crop rotation as well as for animal production; and

(iv) are adopted usually in low-cost structural management systems.

Irrigated pastures call for special attention to matters related to the provision of water (and sometimes its quality), and to the choice of adapted species for specific soil and environmental conditions.

6.2 PLANT COMPONENTS

The recommended mixtures, including seeding rates, for irrigated pasture in the Pacific Coast region of the USA are presented in Table 6.1. White clover, ryegrasses and orchard grass are the usual components except for the southern desert area where bermuda grass is used, sometimes overseeded with annual ryegrass. A common basic mixture for the Great Plains region is smooth bromegrass and orchard grass, with or without alfalfa (Nichols et al., 1976; Anderson and Dunn, 1982). Reed canary grass and Garrison creeping foxtail are both adapted to wet conditions. Recently there has been increasing interest in the use of klein grass for irrigated pasture in southern New Mexico (Fender and Lugg, 1985). Highest yields in central Saskatchewan, Canada, were obtained from meadow bromegrass and orchard grass, followed by timothy, smooth bromegrass and Kentucky bluegrass (Knowles and Sonmor, 1985).

TABLE 6.1

Pasture mixtures for irrigated pasture in the Pacific Coast region. (Adapted from Marble et al., 1985)

Mixture	Seeding rate (kg ha^{-1})	Mixture	Seeding rate (kg ha^{-1})
Higher rainfall, coastal areas			
White clover (cv. Huia)	3	or: Ladino clover	3
Perennial ryegrass	12–15	Salina strawberry clover	2–3
Orchard grass	12–15	Annual ryegrass	2
		Perennial ryegrass	3
		Orchard grass	6
	27–33		16–17
Interior and warmer areas			
Ladino clover	3	or: Alfalfa	5
Orchard grass or	15	Orchard grass	6
Tall fescue	20		
	18–23		11
		or: Ladino clover	3
		Salina strawberry clover	2–3
		Annual ryegrass	2–5
		Perennial ryegrass	3–5
		Orchard grass	6–10
			16–26
Southern interim valleys and salinc soils			
Ladino clover	2	or: Bermuda grass sprigs	...
Narrowleaf trefoil	3		
Annual ryegrass	2		
Perennial ryegrass	3		
Orchard grass	6		
	16		

ment. However, before embarking on such a development, it should be determined if the expected returns from animal production using irrigated pasture are greater than from green chop or conserved forages. On previously non-cultivated land, however, it should be assessed whether soil and other site factors are capable of supporting highly productive pastures without causing undue difficulties in the operation of irrigation equipment (Nichols and Clanton, 1985).

In the historical development of irrigated pastures in southeastern Australia, land was chosen on the basis of topography to ensure easy water distribution and low development costs. Suitability of soils capable of supporting highly productive pasture was often a secondary consideration. This explains the occurrence of extended pasture areas on heavy-textured soils with limited production potential, requiring soil amendments such as gypsum (Greene and Wilson, 1984) or deep ripping to improve water penetration. Some of the better light-textured soils which became waterlogged and saline have been reclaimed by a combination of leaching irrigations and the maintenance of a pasture cover (Myers, 1970; Grieve et al., 1986).

6.4 LAND PREPARATION AND SOWING

Besides seedbed preparation, cultivation serves the purpose of enabling the most economic and uniform application of water for irrigation. On previously irrigated land, cultivation is usually restricted to seedbed preparation, with weed control as a major consideration. The following options should be considered:

(i) growing a hay or grain crop prior to sowing;

(ii) irrigate weeds to encourage their growth and then disc them under early in the seedbed preparation; and

(iii) the use of herbicides.

Sowing of perennial cool-season species should take place in autumn early enough to ensure their establishment before the onset of cold weather and to avoid troublesome summer weeds. Spring sowing is warranted in areas where severe winters may kill autumn-sown seedlings. Warm-season species are preferably sown in spring. Mid-autumn sowings of annual ryegrass in southeastern Queensland gave the highest yields, the least establishment problems, and a more even, weed-free pasture (Lowe et al., 1986). Late summer irrigation and establishment of winter annual legumes in California's Mediterranean range land are discussed by Taggard et al. (1976).

6.5 MANAGEMENT

6.5.1 Irrigation

All of the following techniques are available for water application: wild flooding, border check, contour check, corrugation, furrows, and sprinkler. Choice of the method depends on variables such as topography, soil depth and texture, permeability of soil and subsoil, amount of water available, rate of water delivery, surface drainage, cost of water, and pump size and power. Practical aspects relating to the choice of the system are discussed elsewhere (Peterson et al., 1959; Raguse et al., 1967; Turelle and Austin, 1967; George et al., 1980). Pressurized irrigation equipment is better adapted to pasture irrigation than are methods employing non-pressurized equipment. Sprinklers enable frequent light applications of water which are often required because of the comparatively shallow root system of most pasture species.

Irrigation of annual regenerating pastures in southeastern Australia starts in early autumn and continues until winter rains are adequate. It is resumed in spring until clovers set seed and the stand 'hays off'. Irrigation water requirement, depending on seasonal rainfall, is 300–500 mm year^{-1}. If irrigation is commenced too early, when temperatures are still high (i.e. in late summer), the rate of germination and establishment of both subterranean clover and Wimmera ryegrass are restricted (Kelly and Mason, 1986). Gray and brown-cracking soils, used extensively for irrigated pastures, are reported to require sometimes more than 200 mm of water to fill the cracks at the end of summer. These soils often have a high cation exchange capacity with up to 15% Na^+, causing unstable soil aggregates. Davidson and Quirk (1961) obtained spectacular responses in emergence and subsequent growth of subterranean clover from dissolving gypsum in irrigation water. Perennial pastures are irrigated throughout periods of deficient rainfall and this could amount to 1150–1300 mm year^{-1} of water (Myers, 1970). It was found that waterlogging considerably depressed the yield of perennial ryegrass (Donohue et al., 1984) while reduction of white clover yield is reported by Blaikie and Martin (1987). Drainage of surface irrigation water usually remedied this problem. In contrast, paspalum was much less sensitive to either water shortage or waterlogging (Grieve et al. 1986).

Perennial species are the major components of irrigated pastures in North America. In the Great Plains region, about half of the total water requirement of nearly 900 mm year^{-1} is obtained from irrigation sufficient to maintain adequate soil water in the pasture root zone throughout the growing season (Somerhalder and Clanton, 1972; Dodds et al., 1973).

6.5.2 Fertilization

The use of fertilizers depends on a great number of interacting factors:
(i) are the nutrients needed as basic fertilizers (sward establishment), and/or as top-dressing (sward maintenance, often in terms of grass/legume balance);
(ii) are they applied for short-term duration (usually at high stocking rates), seasonal (because of weather constraints) or year-long grazing;
(iii) is the pasture grazed continuously or rotationally and by what type of animal, with emphasis on forage quality;
(iv) is the pasture exclusively for grazing or is it also harvested by cutting; (iv) is the pasture part of a crop rotation scheme; and, above all
(v) is fertilization economically justified?

In California a considerable number of fertilizer experiments on irrigated pasture were carried out during the 1950s and 60s (Martin et al., 1965). Most soils of the northeastern mountain valleys with native meadow or improved pasture species are markedly deficient in S and often in P, while sometimes B is lacking. Applied P and B tend to remain effective for several years, whereas gypsum or other S sources must be applied more frequently. Spectacular yield responses were obtained after the application of fertilizer N. However, the apparent N recovery decreased sharply and the percentage of clovers decreased with the increase of N fertilizer.

The profitability of a fertilizer program needs careful assessment. A value must be placed on the expected extra forage resulting from fertilization, and on the cost of the fertilizer to bring about the increase in forage. The efficiency of a fertilization program for pasture must also be judged in terms of value of the extra feed when needed for a specific purpose and/or of the expected increase in stocking rate which can be achieved.

Many of the locations with legume-dominant irrigated pasture in the central valley of California are shallow reddish, hardpan and old claypan soils usually deficient in available P and S but with sufficient available K. Correct fertilization usually resulted in increased yields and stimulation of clovers, which in turn provided additional N for the grass component of the pasture mixture. Single applications of N fertilizer remained effective for only a few weeks. Continued N application to a ladino clover–orchard grass mixture increased total forage yield, but orchard grass replaced the clover (Peterson and Bendixen, 1961). Recently Kanyama-Phiri et al. (1990) studied the effect of split amounts of N on yield and species balance of a perennial grass–legume mixture at Davis, California. These split applications were useful in maintaining a desirable grass–legume balance and high DM yield in closely-grazed pastures, but at N levels effective in achieving these objectives, a low recovery of applied N is expected. N and P fertilizers significantly increased forage yields but produced only a small increase in the P or the protein concentration of the forage. These studies showed that the application of N fertilizer on legume-dominant pasture was usually uneconomic.

In the dallis grass and bermuda grass dominant pastures of the hot valleys of southern California, N + P fertilizers showed spectacular yield increases during summer, although the protein content of the forage was not altered appreciably. Beef production of yearling steers from fertilized fields almost doubled compared with unfertilized fields, and animals in the fertilized fields gained at a slightly higher rate.

Nitrogen rates of 200–250 kg ha^{-1}, sometimes applied as liquid urea in center-pivot sprinkler systems, are quoted for maximum response of grass–legumes mixtures in the Great Plains region (Somerhalder and Clanton, 1972).

Top-dressing with N fertilizer on irrigated pasture grazed by dairy cows in eastern Australia has been expanding recently (King and Stockdale, 1980; Stockdale and King, 1980; Chopping et al., 1982a). The main reasons for this are the possibilities of:

(i) increasing the supply of paddock feed, which enhances milk output per ha; and

(ii) reducing the seasonal fluctuations in milk supply, specifically by increasing feed supply during the cool season.

In northern Queensland, Chopping et al. (1982a) showed that N-fertilized pangola grass can be equal to or surpass temperate pasture in levels of milk production. Milk production levels per cow were acceptable though not high, depending on stocking rate. The use of irrigation on tropical pastures has not been widely adopted by the Queensland dairy industry because their production under natural rainfall is usually sufficient in a summer-based rainfall area. However, improved reliability is achieved from supplementary irrigation during dry periods. The tropical grass system does not solve the traditional problem of the marked reduction in pasture supply and quality over the cool dry season.

Irrigated ryegrass and/or clover pastures, when seeded annually in autumn at a high seeding rate, were found to produce between 10 and 17 Mg of milk ha^{-1} during 200 d over the winter–spring period. Ryegrass produced more DM but clovers were of considerably higher value for milk production (Chopping et al., 1983).

The effect of stocking rate and N fertilizer on the productivity of strip-grazed irrigated pastures was studied by Stockdale and King (1980) in

Victoria. Paspalum, perennial ryegrass, cocksfoot and white clover were dominant in these pastures. N fertilizer increased grass production by as much as 20% specifically in spring, but did not reduce the absolute contribution of the clover. Top-dressing with N had no effect on total milk or butterfat production per cow, but consistently increased protein production, and also increased live weight per cow, particularly at the high stocking rates.

6.5.3 Grazing

Three different grazing systems are recognized: continuous (i.e., set stocking), rotation (paddock), and strip grazing. All three systems are applied in irrigated pasture. With continuous grazing, animals graze freely over the entire area. Their selective grazing may cause a reduction in the forage production of the most palatable species and allow the less palatable ones to become more mature, less nutritious and more competitive. In the rotation grazing system the pasture is subdivided into separate paddocks where animals graze for only a certain number of days, allowing a recovery period between grazings. This system improves but does not eliminate selective grazing. Strip grazing restricts the area to the amount rationed for each day of feeding, usually by the use of a portable electric fence, which reduces selective grazing to a minimum. In controlling the size of the grazing strips, forage quality can be optimized, by avoiding either overgrazing or under-utilization.

Any of the three systems can potentially provide optimum performance per animal, depending on energy intake, which is interrelated with the availability of pasture, leafiness of the sward, and digestibility and intake of the forage (Blaser, 1982). Major management options which account for the choice of grazing system are:

(i) whether pasture is the sole source of roughage or is supplemented with other forages and/or concentrates;

(ii) the type of grazing animal;

(iii) the duration of grazing; and

(iv) the cost of irrigation, labor and fencing.

Reports on studies of different grazing management systems are reviewed briefly.

In California, Hull et al. (1967, 1971) investigated the performance of yearling beef steers and heifers grazing continuously or rotationally irrigated pasture composed of perennial temperate species. Continuous yearlong grazing consistently resulted in higher average outputs per ha and per animal than rotational grazing (Table 6.2). The small differences

TABLE 6.2

Average yearly production of beef steers and heifers on irrigated pasture as influenced by grazing system. (Adapted from Hull et al., 1971)

	Grazing system		
	Continuous	Five-field rotation	Two-field rotation
Animal days ha^{-1}	1501	1501	1550
Live weight gain kg $animal^{-1}$ d^{-1}	0.73	0.65	0.68
Live weight gain kg ha^{-1}	1096	976	1057

depended on the type of rotational grazing. The percentage of legumes was consistently higher in the continuously than in the rotationally grazed sward. Regardless of the grazing system, there was a trend toward a higher percentage of grass over time. However, the significance of the botanical shifts in terms of plant nutritive value and productivity apparently were not reflected in animal performance.

George et al. (1992) report of a program of controlled grazing and feed budgeting of irrigated pasture in northern California, which was explored in a number of ranches during a four-year period. This forage budgeting program is based on monitoring of:

(i) the amount of pregrazing herbage mass present in each paddock at the beginning of a rotation;

(ii) the post-grazing herbage mass target;

(iii) expected pasture growth rate;

(iv) forage energy concentration; and

(v) the quantity and quality of supplemented feeds.

Methods for quantifying these parameters are described. The report includes a detailed example of how to combine controlled grazing with adjusted feed budgeting.

Forage quality and palatability in rotationally grazed pastures are bound to decline from the day animals enter one paddock until the day they are moved to the next one, depending on the length of the grazing period. These sequential fluctuations in pasture condition affect the performance of beef cattle to a lesser degree than that of dairy cattle. Hull et al. (1967, 1971) found only a small difference in animal output of beef steers between continuously and rotationally grazed irrigated pasture. A special form of rotational grazing, known as 'short rotation', partly resolves this problem of fluctuating pasture conditions and is particularly effective during short grazing seasons. This system is based on grazing periods usually lasting

et al. 1982b; Goodchild et al. 1982). Major factors affecting the establishment of overseeded species into existing swards are: sward characteristics, seedbed preparation, selection of species, and seeding time and rate.

Characteristics of the existing sward that may determine the success or failure of the overseeding operation are: sward density, rate of autumn growth, susceptibility to cultivation and to shading during dormancy, and onset of growth renewal in the spring. Seedbed preparation using a slightly opened tandem disc implement or sometimes a rotary cultivator is usually a prerequisite for successful germination and seedling establishment. Loosening the soil without ill effects to the existing sward is not always easy to achieve, particularly when the soil surface layer is compacted and/or densely interwoven with roots (so-called 'sodbound') (Kleinig, 1966). Successful shallow cultivation of paspalum has a renovating effect by promoting the decomposition of thatch, manure and plant debris, and stimulates regrowth in the spring. Too rigid cultivation may destroy the plant cover. Overseeding is also useful in that it allows cattle access to grazing in wetter soil situations. The pasture is much more stable if the area has not been fully plowed up for an autumn seeding of temperate species. This is a very important consideration in the subtropical dairying areas of southern Queensland (K.F. Lowe, personal communication).

Crops with small seeds, like berseem, medics and ryegrass, become established more easily than those with large seeds like vetch and oats. The effects of seed rate and size, seedling growth, and competition on the survival of species overseeded into existing swards are reviewed by Cook (1980). Common guidelines of zero or minimum tillage for the establishment of overseeded species, including the use of herbicides for the suppression of summer pasture, are provided by Hawley et al. (1991).

6.6 ANIMAL COMPONENTS

Beef and dairy enterprises are the principal users of irrigated pasture, depending on competitiveness with other irrigated crops such as corn (silage) and alfalfa (hay), with feed grains or horticultural crops. In the USA, irrigated pasture for beef production is utilized mainly as a complementary crop to other forages and not as a primary forage resource. Its specific role in the forage program depends on the associated resources and the livestock program (Wheeler, 1981; Nichols and Clanton, 1985). In Australia, on the other hand, irrigated pasture is mostly a principal feed source for dairy cattle. For increased milk production, however, the feeding of supplemental forages and concentrates is often necessary in order to maximize pasture and milk production efficiency (Radcliffe, 1986). Representative animal production data from irrigated pasture are listed in Table 6.3.

6.6.1 Beef Cattle

In a year-round cow/calf operation in regions of the USA with a Mediterranean climate, available DM, energy and/or protein content from the range is limited during 6–9 months of the year. Irrigated pasture is sometimes used to complement seasonal DM production during that dry period instead of movement of livestock from seasonally deficient areas of forage production. The birth and average daily live weight gains of calves did not differ significantly between a system of yearlong use of cleared range without supplements ('simple system'), and one with cleared and reseeded range plus irrigated pasture and winter supplementation with alfalfa cubes ('complex system') (Raguse et al., 1980). Average cow weights reflected the seasonal changes in forage availability and quality. Economic returns were similar but low in both systems. The area of land required per cow/calf unit was less for the complex system but total investment was higher for the simple one, whereas operation costs, depreciation and interest were higher for the complex system.

In a cow/calf enterprise in eastern North Dakota, a five-year study was made of beef production on irrigated land, comparing dry-lot with pasture (Table 6.3). This environment requires confinement of the beef cow herd for 6–7 months each year. Housing facilities were essentially the same for cattle grazed during the growing season or retained in dry-lot and fed. Although gains of the cow/calf pairs on pasture were slightly higher than in dry-lot, in terms of total cow/calf gain the latter system provided 30–50% more available feed energy per irrigated ha than did the pasture management system. About twice as many cow/calf pairs could be maintained per forage hectare when the dry-lot system was used compared with irrigated pasture. On the other hand, the weaning of calves on irrigated pasture in the autumn and early winter was found to be an excellent practice (Nichols et al., 1976). Reproductive performance of cows on irrigated pasture was better than either in dry-lot or on range (Clanton, 1977).

Grain supplementation on irrigated pasture did not result in significantly improved gains of beef heifers, but enabled a marked increase in stocking rate (Hull and Raguse, 1978; Anderson and Dunn, 1982). This may be a reflection of the narrow N-to-energy ratio or the relatively high level of nonprotein N in irrigated pasture forage (Nichols and Clanton, 1985).

TABLE 6.3

Representative data on animal production from irrigated pasture

Location and year(s)	Pasture species[z]	Grazing system	Supplements (kg $head^{-1}$ d^{-1})	Animals (ha^{-1})	Period (d)	Live weight gain		Reference
						kg $head^{-1}$ d^{-1}	kg ha^{-1}	
YEARLING BEEF STEERS								
North Dakota								
1978	O,SB,A	5-field rotation	2.7 cracked corn	3.6	112	0.43	290	Anderson and Dunn
1979	"	"	2.3 rolled barley	5.9	98	0.70	460	(1982); border irrigation
California								
	PR,O,WC	2-field	None	7.7	200	0.75	1150	Raguse et al. (1975)
	SC	1-week rotation	2.9 alfalfa cubes	12.1	200	0.70	1703	
Oregon								
1973	TF,C	10–14 d rotation	None	7.5	102	0.77	647	Gomm (1979); sprin-
			1.4 barley	7.5	102	0.81	689	kler
Queensland								
1981	AR	3–4 week	None	6.2	109	0.91	1070	Robbins and Bushell
1984	RB	rotation	None	6.7	57	0.51	610	(1987); surface irrigation
BEEF COW/CALF HERD								
North Dakota								
1973–77	O,SB,A	Strip-grazing	None	AUM[y] 18	105	0.75[x]	297	Dunn and Olson (1978); central pivot

California 1973–77	O,PS,PR, WC	Rotational	None	AUM 10.3	180	0.76[x]	70[x]	Raguse et al. (1980); sprinkler
DAIRY COWS						Milk production		
						kg cow^{-1}	kg ha^{-1}	
Queensland 1971–73	PA	Rotational	None	7.9	273	2,513	19,851	Chopping et al. (1982); surface irrigation
			3.6 molasses	7.9	283	3,185	25,164	
Queensland 1975	AR	Strip-grazing	1.8 sorghum	7.6	207	2,731	20,890	Robbins et al. (1980); surface irrigation
Victoria 1973–74	PS,PR,O, WC	Strip-grazing	Hay[w]	8.6	Year round	2,081	17,900 (847 butter-fat)	King and Stockdale (1980); Stockdale and King (1980); surface irrigation

[z]O: orchard grass; SB: smooth bromegrass; A: alfalfa; PA: pangola grass; AR: annual ryegrass; C: clover; PS: paspalum; PR: perennial ryegrass; WC: white clover; CF: creeping foxtail; SC: strawberry clover; TF: tall fescue; RB: annual ryegrass/berseem clover mixture.
[y]AUM: Animal Unit Month.
[x]Calves only.
[w]Pasture-conserved hay during dry period.

The adoption of high-forage programs in beef enterprises appears to be economically attractive only at times of increased costs of feed grains and other feeds or of low calf prices in relation to yearling prices.

6.6.2 Dairy Cattle

The use of irrigated pasture for dairying is common practice in Australia, which has a grazing-based industry characterized by low labor and capital inputs. The average yield of dairy products per ha in Victoria and southern New South Wales used to be low, viz., 333 kg butterfat per ha (Habgood et al. 1979). This low animal productivity can be explained partly as the result of:

(i) low forage quality, specifically of the paspalum-dominant pastures, and

(ii) the lack of flexibility in the pasture supply during the irrigation season.

Pasture utilization often does not exceed 50% (Stockdale, 1983; Mason et al., 1987). However, in recent years milk production per cow in Victoria increased significantly together with a decrease in dairy farms and an increase of the average herd size. This improvement in productivity is mainly the result of greater rationalization in the dairy industry (Greenaway, 1992).

For example, the development of new production systems showed that high levels of efficiency were achieved only by severely limiting pasture supply per animal. Maximizing production per hectare was realized only by increasing stocking rate, through which pasture intake per animal was severely limited (Figs. 6.2 and 6.3). Based on these and other findings, Stockdale and King concluded that: "Increasing pasture utilization will be difficult unless we can at least partly break the link between individual animal performance and rates of pasture utilization. Feeding supplements may provide the answer. A system of heavy grazing pressure to increase pasture utilization plus supplementary feeding to increase individual animal performance could dramatically increase the productivity and profitability of the animal industries in the irrigation areas". This system, known as the Kyabram Dairy System, seeks to utilize the high nutritive value of clover pastures which enables cows to graze for as little as 2 h each day, and the high DM yields of forage corn as silage (Moran et al., 1986; Mason et al., 1987; Moran and Wamungai, 1988; Pritchard et al., 1989).

Irrigated N-fertilized tropical pastures in central Queensland supply 50–100% more milk per year compared with non-irrigated pastures (Chopping et al., 1983). Milk production levels from cows grazing irrigated pastures are

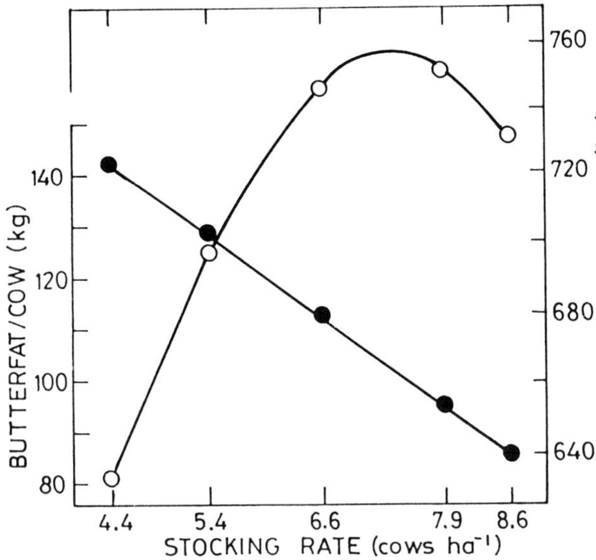

Fig. 6.2. Effect of stocking rate on butterfat production per cow (O) and per hectare (●) for dairy cows grazing irrigated perennial pasture in northern Victoria. (From King and Stockdale, 1980).

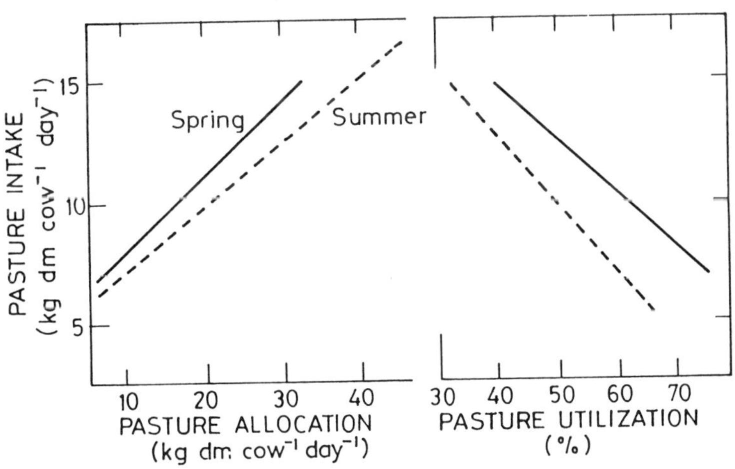

Fig. 6.3. Effect of pasture allocation per cow on pasture intake (left), and of pasture intake on percentage pasture utilization (right) for dairy cows grazing perennial irrigated pasture in northern Victoria. (From Stockdale and King, 1980).

acceptable though not high (Table 6.3). The major nutritional problem facing dairy farms, however, is the marked decline of feed from pastures in the late autumn–winter–spring period because of a dry spell and cool temperatures. Lately, more emphasis has been placed on the provision of stable milk production all year round, in order to meet the market's milk requirements. A substantial increase in the feed supply was obtained by the use of irrigated annually sown ryegrass and/or clover, through which the supply of whole milk in the critical winter–spring period could be maintained (Busby and Van Beek, 1982; Chopping et al., 1982b). Similar results are reported from southeast Queensland (Robbins et al., 1980; Robbins and Bushell, 1987; Lowe and Hamilton, 1985; Moss et al., 1985).

In a recent publication Teese (1992) discusses the economic aspects of forage production and utilization in Australian dairy production systems. To help establish the importance of researching how to improve the efficiency of forage production, it is necessary to consider the costs of alternative sources of forage/feed. The author listed the following options:

(i) upgrading pasture species;
(ii) using N fertilizers;
(iii) increasing the amount of P/K/S trace element fertilizers;
(iv) improved drainage;
(v) using supplementary irrigation on dry-land farms;
(vi) buying/leasing extra land for forage production;
(vii) growing forage crops;
(viii) buying in forage crops such as corn, silage or pasture hay;
(ix) replacing part of the forage ration with grain based concentrates;
(x) adjustment for non-milking stock.

6.6.3 Sheep

In the southern part of California, grazing sheep on irrigated alfalfa is widely accepted to accommodate the early-autumn lambing operation. Sheep producers commonly establish enclosed paddocks by daily erecting temporary fences, finally including entire fields holding flocks of 600 or more sheep. Grazing may result in almost total defoliation of the alfalfa plant, but grazing caused no loss of subsequent forage production nor did it affect the bulk density of the soil. Ponding on the field or grazing under very muddy conditions, however, should be avoided (Pelton et al., 1988). In studies with lambs grazing desert alfalfa during winter it was found that profitable gains were obtained if lambs were moved to fresh pasture after about 10 d of grazing, when the alfalfa plants were nearly devoid of leaves (Guerrero and Marble, 1987; Guerrero et al., 1989).

Highly productive white clover and nondormant alfalfa cultivars were recently found to have potential for improving year-round lamb production in southern New South Wales. Intensive prime lamb production is seen as an alternative to rice production on farms affected by rising water tables. Furthermore, vast areas of dry-land annual pasture can be more exploited efficiently by sowing relatively small areas of irrigated pasture with highly productive complementary feed during the summer–autumn period. The number of wethers carried on a dry-land property could be increased almost five-fold after approximately 3% of the area was seeded with white clover and irrigated (Dawe and Lattimore, 1987). The suckling lamb industry in New South Wales is also exploring the possibilities of using irrigation water on pure-legume pastures in order to cover the late pregnancy and lactation periods of ewes, supplementing the feed from rain-fed pastures (Mason et al. 1987).

6.7 CONCLUSIONS

Justification for the allocation of irrigation water to pasture is usually based on one or more of the following considerations:

(i) greater net returns from animal production by grazing than by feeding cut fodders or from growing cash crops;

(ii) the use of less productive land;

(iii) the use of low quality water;

(iv) providing an effective ground cover for easily-erodible soil or steep slopes;

(v) prolonging the effective growing period during dry seasons;

(vi) supplementing dry annual range;

(vii) low labor requirements and convenience of operation.

Pasture management systems, therefore, are expected to differ greatly among themselves depending on social, economic and environmental conditions, and animal production potential.

A relatively great number of plant species differing in growth season, digestibility, life span, and salt and drought resistance is available, and sometimes used in mixture, for year-round production from irrigated pasture. Recently, however, preference has been given to seeding separate stands of single species or simple mixtures in order to facilitate management practices such as fertilization, and thereby increase the efficiency of utilization.

Soil constraints in terms of permeability are often a primary cause of low pasture productivity. Soil amelioration by deep ripping and the application of gypsum may be useful, but additional techniques must be developed.

Dry-land annual pastures can be better utilized and stock numbers increased when sheep are provided with a small area of highly productive irrigated summer–autumn feed. Also intensive prime lamb production will require year-round pasture systems.

Forage utilization by grazing of species such as dairy cattle is relatively low at present, because often the seasonal pattern of production does not match animal demands. Limiting the amount of pasture offered to grazing animals may be a useful tool to achieve high animal production per unit of land. Conservation of high-yielding fodder crops of high quality — such as corn silage — as supplementary feed to irrigated pasture, is considered an essential element of intensive animal production systems.

Controlled grazing and feed budgeting develop from greater awareness in improving pasture management. The application of more advanced irrigation techniques which enable better control of timing, amount and depth of irrigation should greatly increase water use efficiency.

REFERENCES

Anderson, V.L. and Dunn, B.H. 1982. Grain supplementation of yearling steers grazing irrigated pasture. North Dakota Farm Res. 40: 10–12.

Arnon, I. 1956. Theory and Practice of Field Crops Growing. Sifriyat Hassadeh, Tel Aviv, Israel. 681 pp. (in Hebrew).

Blaikie, S.J. and Martin, F.M. 1987. Limits to the productivity of irrigated pasture in Southeast Australia, pp. 119–122. In: J.L. Wheeler, C.T. Pearson and G.E. Roberts, (Eds.) Temperate Pastures: The Production, Use and Management, Australian Wool Corporation.

Blaser, R.E. 1982. Integrated pasture and animal management. Trop. Grassl. 16: 9–24.

Bokhari, V.G., Alyaresh, F. and Al-Noori, M. 1988. Potentials of forage crops in Saudi Arabia. Arab Gulf J. Sci. Res. Agric. Biol. Sci. 86: 359–367.

Busby, G. and Beek, P.G.H. van. 1982. The economics of pasture based feeding systems. Proc. Aust. Soc. Anim. Prod. 14: 112–115.

Chopping, G.D., Lowe, K.F. and Clarke, L.G. 1983. Irrigation systems, pp. 109–120. In: Dairy management in the 80's: Focus on Feeding. Seminar sessions and farm feeding workshops (Gympie, Qld., Australia). Dep. of Primary Industries, Brisbane, Australia.

Chopping, G.D., Murray, A.J. and Bird, A.C. 1982a. Irrigated pasture systems for milk production in Queensland. Proc. Aust. Soc. Anim. Prod. 14: 106–110.

Chopping, G.D., Thurban, P.N., Moss, R.J. and Stephenson, H. 1982b. Winter–spring milk production responses from the annual autumn oversowing of irrigated tropical pastures with ryegrass and clovers. Proc. Aust. Soc. Anim. Prod. 14: 421–424.

Clanton, D.C. 1977. Finishing cattle on pasture and other forages: Irrigated pasture. J. Anim. Sci. 44: 908–912.

Cook, S.J. 1980. Establishing pasture species in existing swards: a review. Trop. Grassl. 14: 181–187.

Davidson, J.R. and Quirk, J.P. 1961. The influence of dissolved gypsum on pasture establishment on irrigated sodic clays. Aust. J. Agric. Res. 12: 100–110.

Dawe, S.T. and Lattimore, M.E. 1987. Integrated irrigated pasture systems for southern New South Wales. pp. 495–497. In: J.L. Wheeler, C.T. Pearson and G.E. Roberts, (Eds.) Temperate Pastures: The Production, Use and Management. Australian Wool Corporation.

Dodds, D.L., Johnson, L. and Soiseth, C. 1973. Beef production of irrigated pasture in North Dakota. North Dakota Farm Res. 30 (3): 21–27.

Donohue, G.I., Greene, R.S.B., Willoughby, P. and Wilson, I.B. 1984. The effect of water logging on flood irrigated perennial ryegrass. pp. 85–92. In: W.A. Muirhead and E. Humphreys (Eds.) Root Zone Limitation to Crop Production in Dry Soils, Proc. Australian Soc. Soil Sci., (Griffith, NSW, Australia).

Dovrat, A. and Arnon, I. 1971. Pattern of intensive roughage production from irrigated forage and pasture crops in Israel. Bull. Recher. Agron. Gembloux. H.S. 601–605.

Dovrat, A. and Kipnis, T. 1983. Maximum production from irrigated rhodes grass and green panic in a Mediterranean environment. pp. 574–576. In: Proc. 14th Int. Grassland Congr. (Lexington, KY).

Dunn, B.H. and Olson, H.M. 1978. Cow-calf beef production with irrigated forages. North Dakota Farm Res. 36 (1): 13–21.

Evans, T.R. 1982. Overcoming nutritional limitations through pasture management. pp. 344–357. In: J.B. Hacker (Ed.) Nutritional Limits to Animal Production from Pastures. CAB. Farnway Royal, UK.

Fender, S.E. and Lugg, D.G. 1985. Kleingrass production and nitrogen percentage under different nitrogen fertilization and irrigation levels. Forage and Grassland Conf. (Lexington, KY) pp. 183–185.

George, M.R., Clawson, J., Weitkamp, B. and Willoughby, B. 1980. Management of small pastures. Div. Agric. Sci. Univ. Calif. Leafl. 2906. 8 pp.

George, M.R., Robbins, M.E., Bell, F.L., Riet, W. van, Markegard, G., Lile, D.F., Wilson, C.B. and Barr, Q.J. 1992. Managers control of forage levels and animal performance. Cal. Agr. 46 (3): 27–30.

Gomm, F.B. 1979. Irrigated pasture for range livestock. Oregon Agric. Exp. Stn. Bull. No. 635. 15 pp.

Goodchild, I.K., Thurbon, P.N., Sibbick, R. and Shepherd, R. 1982. Effect of land preparation and nitrogen fertilizer on yield and quality of temperate species introduced into a tropical grass sward during autumn. Aust. J. Exp. Agric. Anim. Husb. 22: 88–94.

Greenaway, R. 1992. Trends in dairy farm profitability. In: Smith, G.H. and Capicchiano, P.J. (Eds.) Dairying for the Future. Proc. Australian Soc. of Anim. Production. Melbourne, Australia.

Greene, R.S.B. and Wilson, I.B. 1984. Some physico-chemical properties of an exposed B horizon of a red-brown earth treated with gypsum. Aust. J. Soil Res. 22: 357–363.

Grieve, A.M., Dunford, E., Marston, D., Martin, R.E. and Slavich, P. 1986. Effects of water logging and soil salinity on irrigated agriculture in the Murray Valley: a review. Aust. J. Exp. Agric. 26: 761–777.

Guerrero, J. and Marble, V.L. 1987. Alfalfa growing strategies to maximize lamb gains. Proc. 17th Calif. Alfalfa Symp. (El Centro, CA) pp. 77–80.

Guerrero, J., Marble, V.L. and Hernandez, C. 1989. Desert alfalfa grazing — Its the leaf that counts. Calif. Agr. 43: 31–32.
Habgood, R., Turnbull, E. and Farrow, S. 1979. Dairy farm management study 1978/1979. Govt. of Victoria, Australia, Res. Proj. Ser. 72. 35 pp.
Hawley, M., Lowe, K., Launders, T. and Moore, K. 1991. Temperate pastures and forage crops from Gympie to Casino. Queensl. Dep. Primary Ind. Inf. Ser. QI 91001. pp. 30–34.
Heath, M.E., Barnes, R.F. and Metcalfe, D.S. 1985 (Eds.). Forages. 4th ed. Iowa State Univ. Press, Ames, IA.
Heerden, J.M. van and Tainton, N.M. 1988. Influence of grazing management on the production of an irrigated grass/legume pasture in the Ruens area of the Southern Cape. J. Grassl. Soc. South Afr. 5: 130–137.
Hull, J.L. and Raguse, C.A. 1978. Irrigated pasture for beef heifers to be bred as yearlings. J. Anim. Sci. 46: 878–883.
Hull, J.L., Meyer, J.H. and Raguse, C.A. 1967. Rotation and continuous grazing on irrigated pasture using beef cattle. J. Anim. Sci. 26: 1160–1164.
Hull, J.L., Raguse, C.A. and Henderson, D.W. 1971. Further studies on continuous and rotational grazing of irrigated pasture by yearling beef steers and heifers. J. Anim. Sci. 32: 984–988.
Jones, C.A. 1985. C_4 Grasses and Cereals: Growth, Development and Stress Response. John Wiley & Sons, New York, NY. 419 pp.
Jones, R.J., Griffith, D.J., Wiate, R.B. and Fergus, I.F. 1968. The production and persistence of grazed irrigated pasture mixtures in Southeastern Queensland. Aust. J. Exp. Agric. Anim. Husb. 8: 177–189.
Kanyama-Phiri, G.Y., Raguse, C.A. and Taggard, K.L. 1990. Responses of perennial grass–legume mixture to applied nitrogen and differing soil textures. Agron. J. 82: 488–495.
Kelly, K.B. and Mason, W.K. 1986. Evaluation of some annual *Trifolium* spp. under irrigation. Austr. J. Exp. Agric. 26: 79–85.
King, K.R. and Stockdale, C.R. 1980. The effects of stocking rate and nitrogen fertilizers on the productivity of perennial pasture grazed by dairy cows. J. Exp. Agric. Anim. Husb. 20: 537–542.
Kipnis, T., Feigin, A., Vaisman, I. and Shalhevet, J. 1983. Waste-water application, dry-matter production, and nitrogen balance of rhodesgrass grown on fine-textured soil or on sand dunes. 14th Int. Grassland Congr. (Lexington, KY), pp. 288–290.
Kleinig, C.R. 1966. Mats of unincorporated organic matter under irrigated pastures. Aust. J. Agric. Res. 17: 327–333.
Kleinschmidt, F.M. and Skerman, P.J. 1977. Irrigation of tropical pasture legumes. pp. 150–158. In: Tropical Forage Legumes. FAO Plant Prod. Prot. Ser. 2, pp. 150–158.
Knowles, R.P. and Sonmor, L.G. 1985. Grasses for irrigated pasture in central Saskatchewan. Can. J. Pl. Sci. 65: 455–458.
Lowe, K.F. 1978. Irrigated pasture research in southeast Queensland. pp. 52–73. In: Management of Southeast Queensland Coastal Pastures, Queensl. Dep. Prim. Ind. Brisbane, Australia.
Lowe, K.F., Bowdler, T.M. and Mulder, J.R. 1986. Effect of early sowing on establishment, growth pattern and rust infection of annual sowings of ryegrass (*Lolium*

spp.) in sub-coastal southeast Queensland. Trop. Grassl. 20: 113–120.

Lowe, K.F. and Hamilton, B.A. 1985. Dairy pastures in the Australian tropics and subtropics. pp. 68–79. In: G.J. Murtagh and R.M. Jones (Eds.) Proc. 3rd Austr. Conf. Trop. Pastures. Trop. Grassland Soc. Australia Occ. Publ. 3.

Marble, V.L., Raguse, C.A., McGuire, W.S. and Hannaway, D.B. 1985. Hay-pasture seedings for the Pacific Coast States. pp. 400-410. In: M.E. Heath, R.F. Barnes and D.S. Metcalfe (Eds.) Forages. 4th ed., Iowa State Univ. Press. Ames, IA.

Martin, W.E., Rendig, V.V., Haig, A.D. and Berry, L.J. 1965. Fertilization of irrigated pasture and forage crops in California. Calif. Agric. Exp. Sta. Bull. 815. 35 pp.

Mason, W., Kelly, K., Blaikie, S. and Stockdale, R. 1987. New directions for irrigated pastures. Proc. 4th Australian Agriculture Conf. (Melbourne, Australia), pp. 100–117.

Moran, J.B., Stockdale, C.R. and Trigg, T.E. 1986. Maize silage as a supplement for grazing dairy cows. Proc. Aust. Soc. Anim. Prod. (Canberra, Australia) 16: 283–286.

Moran, J.B. and Wamungai, W.N. 1988. Developments in the Kyabram dairy system. Proc. Austr. Soc. Anim. Prod. (Sydney, Australia) 17: 443.

Moss, R.J., Lowe, K., Chopping, G.D., Reason, G.K., Bowdler, T., Bird, A.C., McKeogh, P. and Duncalfe, F. 1985. Effect of stocking rate on milk production from irrigated annual temperate pastures in southern Queensland. Dairy Prod. Conf. Austr. Soc. Anim. Prod. (Albury). pp. 73–74.

Myers, L.F. 1970. Water and irrigated pastures. pp. 260–270. In: R.Milton Moore (Ed.) Australian Grasslands. Australian Nation. Univ. Press, Canberra, Australia.

Nichols, J.T. and Clanton, D.C. 1985. Irrigated pastures. pp. 507–516. In: M.E. Heath, R.F. Barnes and D.S. Metcalfe (Eds.) Forages. 4th ed. Iowa State Univ. Press, Ames, IA.

Nichols, J.T., Gray, A.M. Clanton, D.C. and Myran, D. 1982. Alfalfa, cicer milvetch and cool-season grasses. Irrigated pasture — animal production. Nebr. Agric. Exp. Stn. Beef Rep. MP43: 31–32.

Nichols, J.T., Moore, M.J., Hoaston, L.F. and Clanton, D.C. 1976. Grasses for irrigated pastures. Nebraska Beef Cattle Rep. EC 12: 76–218.

O'Grady, R. and Cassidy, G.J. 1976. An irrigated pasture system based on annual ryegrass sowings. Queensland Agric. J. 102: 3–9.

Pelton, R.E., Marble, V.L., Wildman, W.E. and Peterson, G. 1988. Fall grazing by sheep on alfalfa. Calif. Agric. Sept.–Oct.: 4–5.

Peterson, M.L. and Bendixen, L.E. 1961. Plant competition in relation to nitrogen economy. Agron. J. 53: 45–48.

Peterson, M.L., Osterli, V.P. and Berry, L.J. 1959. Managing irrigated pastures. Calif. Agric. Exp. Stn. Circ. No. 476. 31 pp.

Pritchard, K.E., Mason, W.K., Stockdale, C.R. and Moran, J.B. 1989. Integrating forage maize with pasture legumes for efficient dairy production in south eastern Australia. Proc. 16th Int. Grassland. Congr. (Nice, France) 16: 1143–1144.

Radcliffe, J.C. 1986. Dairying. pp. 131–147. In: G. Alexander, O.B. Williams (Eds.). The Pastoral Industries of Australia. Sydney Univ. Press, Sydney, Australia.

Raguse, C.A., Berry, J.L. and Street, J.E. 1967. Irrigated pastures in California. Calif. Agric. Exp. Stn. Ext. Serv. Circ. 545. 35 pp.

Raguse, C.A., Hull, J.L., Henderson, D.W. and Osman, A. 1975. Perennial irrigated pastures. II. Average daily gain and carcass characteristics of yearling beef steers on

irrigated pasture supplemented by alfalfa cubes. Agron. J. 67: 580–582.
Raguse, C.A., Hull, J.L. and Delmas, R.E. 1980. Perennial irrigated pastures. III. Beef calf production from irrigated pasture and winter annual range. Agron. J. 72: 493–499.
Raguse, C.A., Taggard, K.L., Hull, J.L., Daley, C.A. and Connor, J.M. 1989. Short duration grazing on irrigated pasture. Calif. Agric. 43: 4–7.
Robbins, G.B., Busby, G.J. and Faulkner, G.B. 1980. Animal production from annual pastures of irrigated ryegrass in the Burnett district of south-east Queensland. Trop. Grassl. 14: 40–43.
Robbins, G.B. and Bushell, J.J. 1987. Productivity of grazed irrigated pastures of ryegrass, berseem clover, and ryegrass/berseem clover in south-east Queensland. Trop. Grassl. 21: 133–138.
Skerman, P.J. and Riveros, F. 1990. Tropical Grasses. FAO Plant Prod. Prot. Ser. 23. 832 pp.
Smith, R.C.G., Mason, W.K., Meyer, W.S. and Barrs, H.D. 1983. Irrigation in Australia: Development and prospects. Adv. Irrig. 2: 99–153. Academic Press, New York, NY.
Somerhalder, B.R. and Clanton, D.C. 1972. Pasture irrigation with a center pivot sprinkler system. Trans. Am. Soc. Agric. Eng. 15: 902–904.
Stobbs, T.H. 1977. Milk production, milk composition, rate of milking and animal grazing behaviour of dairy cows grazing two tropical grass pastures under a leader and follower system. Aust. J. Exp. Agric. Anim. Husb. 18: 5–11.
Stockdale, C.R. 1983. Irrigated pasture productivity and its variability in the Shepperton region in northern Victoria. Aust. J. Exp. Agric. Anim. Husb. 23: 131–139.
Stockdale, C.R. 1985. Influence of some sward characteristics on the consumption of irrigated pastures grazed by lactating dairy cattle. Grass Forage Sci. 40: 31–39.
Stockdale, C.R. and King, K.R. 1980. The effects of grazing rate and nitrogen fertilizer on the productivity of irrigated perennial pasture grazed by dairy cows. I. Pasture production, utilization and composition. Aust. J. Exp. Agric. Anim. Husb. 20: 529–536.
Taggard, K.L., Delmas, R.E. and Raguse, C.A. 1976. Late summer irrigation and establishment of winter annual legumes in a Mediterranean type climate. Agron. J. 68: 674–677.
Taggard, K.L., Raguse, C.A., George, M.R., Hull, J.L., Daley, C. and Connor, J.M. 1990. Update on short duration grazing study on irrigated pasture. Calif. Agric. 44 (2): 8–11.
Teese, I. 1992. Economic forage production and utilization. pp. 86–93. In: B. Bartsch and W. Mason (Eds.) Feedbase 2000, CSIRO and Dairy Res. Devel. Corp., Melbourne, Australia.
Turelle, J.W. and Austin, W.W. 1967. Irrigated pastures for forage production and soil conservation in the West. U.S. Dep. Agric. Farmers' Bull. No. 2230. 22 pp.
Wheeler, J.L. 1981. Complementing grassland with forage crops. pp. 239–290. In: F.H.W. Morley (Ed.) Grazing Animals. Elsevier Science Publishers, Amsterdam, The Netherlands.
Whiteman, P.C. 1980. Tropical Pasture Science. Oxford University Press, Oxford, UK. 392 pp.

Chapter 7

FORAGE CORN

7.1 INTRODUCTION

Most management studies in the USA are focused on grain production rather than on silage production. The utilization of corn either for grain or for silage is barely taken into consideration when hybrids are selected by the grower for the latter purpose. The use of hybrids which combine high whole-crop yield with favorable quality characteristics for ensilage has only recently come into consideration in dairy and beef operations. Information on genetic and management factors which affect yield and quality in forage corn has become available mostly from investigations conducted in northern Europe, where corn is grown almost exclusively grown for silage.

Corn and its development are reviewed extensively in a monograph by Sprague and Dudley (1988), which also comprises aspects of diseases (Smith and White, 1988) and pests (Dicke and Guthrie (1988) affecting corn. A compendium of corn diseases is provided by Shurtleff (1988).

7.2 DEVELOPMENT AND GROWTH

Detailed descriptions of the developmental stages and growth of corn are provided by Duncan (1975), Doorenbos and Kassam (1979), and Ritchie et al. (1992). The optimum temperature range for growth is 22–30°C (Shaw, 1988). Developmental stages, shown in Figure 7.1, are summarized briefly.

7.2.1 Canopy Structure

Juvenile Stage

The length of the period from seedling emergence and initial leaf production, until attainment of a constant rate of growth, depends mainly on temperature and plant density. In a Mediterranean climate, the length of

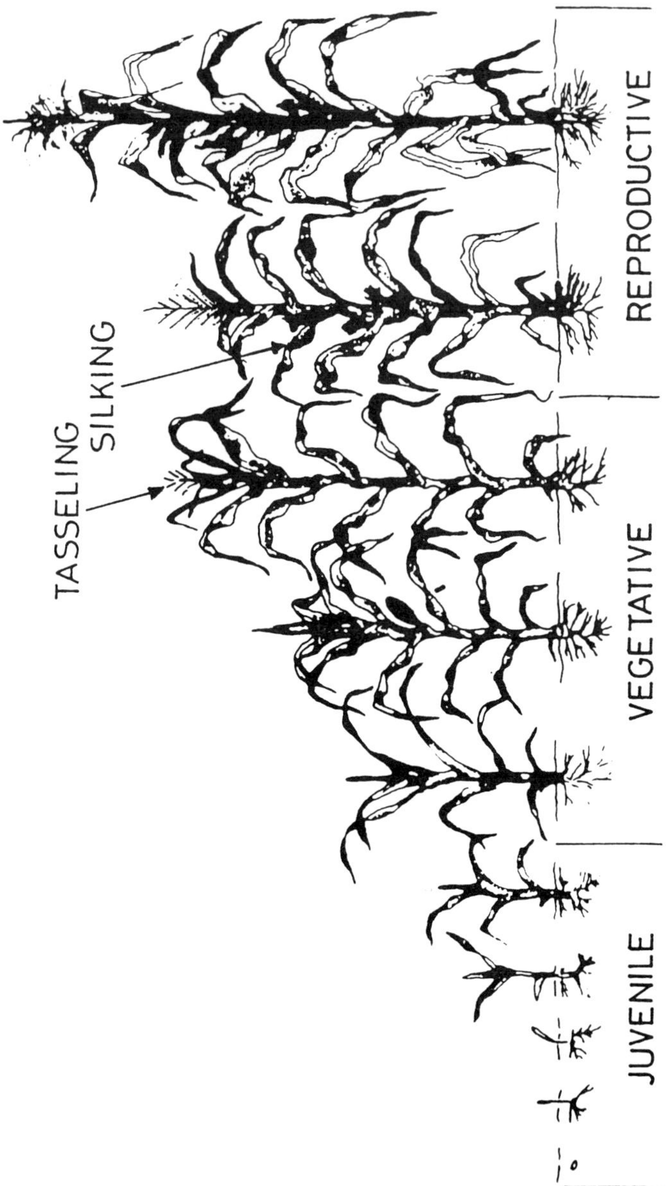

Fig. 7.1. Developmental stages of corn. (From Doorenbos and Kassam, 1979).

the juvenile stage of spring-seeded corn is usually longer than that of a mid-summer seeding because of lower temperatures. In spite of the slower initial growth of spring-seeded corn, the final DM yield is usually higher than that of corn seeded in summer. Exponential growth during the initial growth period continues until the LAI is approximately 3, at which point most of the incoming light is intercepted by the canopy. This stage is reached sooner at high than at low plant density (Eik and Hanway, 1966; Williams et al., 1968; Leshem, 1975).

Vegetative Stage

Development is characterized by further leaf production, increase of LAI, stem elongation, and the initiation of tassel and ear shoots. Ear shoots (potential ears) are initiated just before tassel formation, during which the number of ovules (immature kernels) on each ear and the size of the ear are determined. The rate of DM accumulation during the vegetative stage is usually linear, provided that adequate soil water and nutrients are available. Development of adequate leaf area for the interception of incident radiation has been shown to be closely related to the amount of grain in the whole-crop yield (Williams et al., 1968; Eik and Hanway, 1966). The vegetative stage ends when the last leaf produced is fully expanded and the tassel emerges, thus signals the start of the reproductive stage.

Reproductive Stage

This stage starts when silks, which are a prerequisite for pollination, emerge from the ears. Pollen is shed from the tassel onto receptive silks, resulting in the fertilization of the ovules in the ear. Silk emergence, however, may be delayed as a result of environmental stresses such as a water deficit, or put out of phase with pollen-shedding at high plant densities. In either case, kernel yield will be reduced (Bunting, 1978). Following fertilization, the partitioning of assimilates is directed mainly towards accumulation of DM in the ears, specifically in the kernels. Although maximum kernel yields are not necessarily required when corn is grown for forage, the weight of ears with filled kernels will most likely determine the nutritive value of the crop (Section 7.6).

Leshem et al. (1985) measured the course of DM accumulation during the ear-filling period (Fig. 7.2A). Allocation of assimilates to the ears proceeded asymptotically and the ear fraction made up approximately half of the total yield (i.e., ear/stover ratio equalled almost one) at the end of 30 d, when kernels reached the early dough stage. The ear fraction, contains besides kernels, ca. 22% cobs and husks (see Table 7.4). The DM concen-

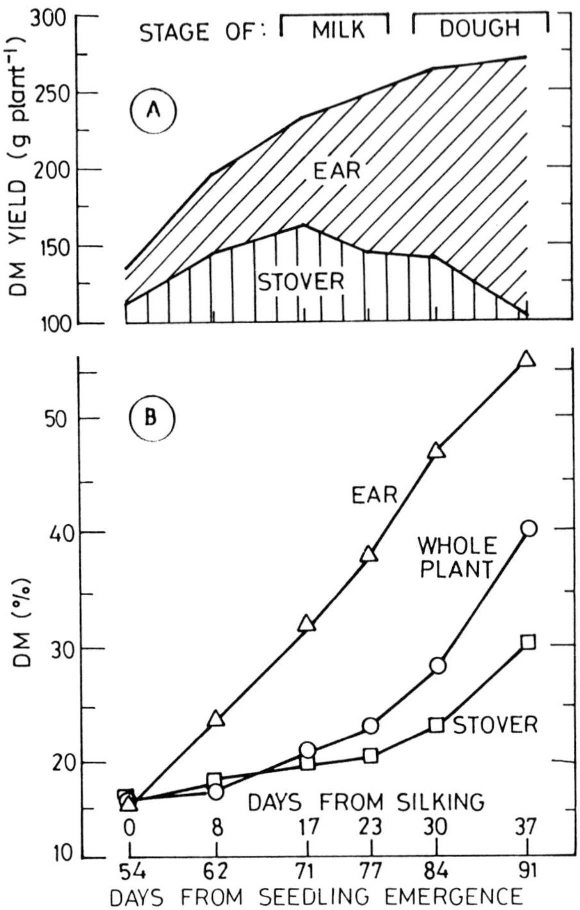

Fig. 7.2. Dry matter accumulation (A) and DM percentage (B) of yield components in a corn plant during the ear-filling period of sprinkler-irrigated forage corn cv. Hason. Plant density = 11 m^{-2}. Bet Dagan, Israel. (Adapted from Leshem et al., 1985).

tration of ears at harvest on day 37 was almost twice that of stover, which equalled ca. 40% DM of the whole-crop — a desirable level for ensiling (Fig. 7.2B.) Kernel filling is characterized by the accelerated accumulation of starch with a steady moisture decrease in the kernels through the ripening stages: milk, dough and dent (ca. 80%, 70% and 55% moisture, respectively).

The rate of vegetative and reproductive development is affected by environmental conditions — especially water supply and nutrient supply and temperature — and by genotype. Numerous investigations showed that the heat requirement of plants from seedling emergence to silking, or from silking to maturity, can be estimated by the direct summation of

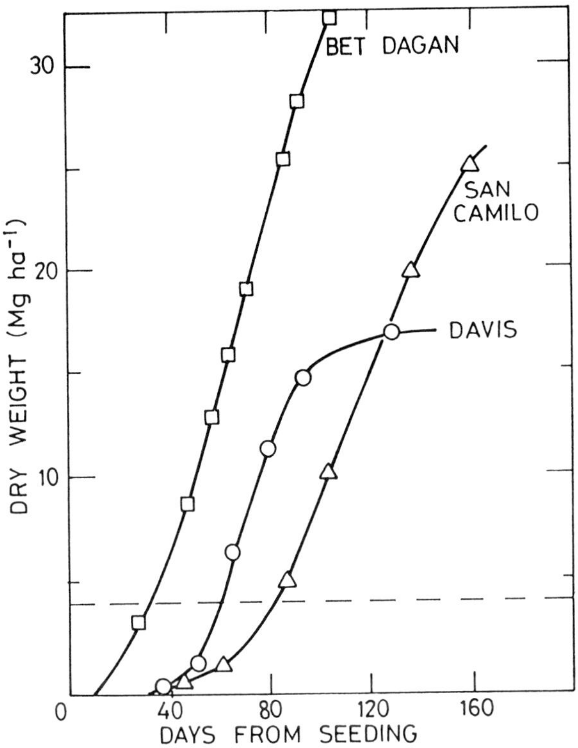

Fig. 7.3. Dry matter accumulation of irrigated forage corn in different environments: Bet Dagan, Israel (Yanuka et al., 1982); Davis, Calif. (Williams et al., 1968); San Camilo, Peru (Versteeg, 1985). (Adapted from Alberda, 1984).

average daily temperatures, with a threshold temperature of 10°C (Tollenaar et al., 1979; Warrington and Kanemasu, 1983; Grant, 1989). Genotype differences between hybrids, i.e., early and late-maturing types, are often the result of differential daylength responses to the date of silking.

7.2.2 Crop Growth Rate

Although different hybrids in different geographic regions showed similar growth rates during the linear phase of growth, when adequately supplied with water and nutrients, final DM yields differed. For example, whole-crop yields were 31, 26 and 20 Mg ha^{-1} at Bet Dagan (Israel), San Camilo (Peru) and Davis (California), respectively, although the rate of linear growth of these crops was quite similar (ca. 400 kg DM ha^{-1} d^{-1}, Fig. 7.3). The number of days required by the San Camilo crop to reach the onset of linear growth (ca. 4 Mg dry weight ha^{-1}) was more than twice that of the Bet Dagan crop. This difference was explained on the basis of lower temperatures at San Camilo, which caused a delay in the rate of juvenile development (Alberda, 1984). Silking of both the Bet Dagan and Davis crops occurred ca. 60 d after seeding. However, the higher prevailing summer temperatures at Davis shortened the ear-filling period and leaf area duration, resulting in lower final whole-crop dry weights than at Bet Dagan. Thus, other growing conditions equal, phenological events mainly depending on temperature, can greatly affect total DM accumulation.

Plant Density

The appropriate plant density for maximum silage production is usually greater than for maximum grain production, because yield–plant population relationships are parabolic for grain but asymptotic for DM production. Whole-crop DM yields increase with plant density to a maximum, with yields remaining more or less constant with further increases in density (Williams et al., 1968; Bunting, 1971; Tetio-Kagho and Gardner, 1988a,b). Morphological and physiological features usually associated with an increase in plant density are:

(i) increase in plant height and in LAI until tasseling (Fairey, 1982);
(ii) decrease in individual plant weight and stem diameter;
(iii) increased incidence of lodging;
(iv) increase in whole-crop yield;
(v) decrease in the ear/stover ratio (due to barrenness or unfilled ears), depending on genotype.

7.3 ESTABLISHMENT

7.3.1 Soil Requirements

Forage corn is cultivated on soils of widely different physical and chemical properties including soil depth. Within these wide ranges, however, there is a corresponding variation in the crop's productivity. Most ideal are intermediate textures of loam to silt loam in the surface horizon, and a somewhat higher clay content — such as silt loam — to silty clay loam in the subsoil. This combination, complemented with good structural properties, allows good storage of water and nutrients and a degree of permeability favorable to water intake and air exchange. Poorly drained soils restrict proper root development because of the lack of aeration (van Schilfgaarde, 1974; Olson and Sander, 1988). In addition, corn is sensitive to a number of herbicides and the possible presence of their residues in the soil should be carefully checked.

Tillage practices and seed-drilling equipment are reviewed by Olson and Sander (1988).

7.3.2 Methods and Date of Seeding

Seeds of corn germinate rapidly when the soil temperature at a depth of 7–10 cm reaches 15–16°C. Early-spring seedings are sown at a depth of ca. 3 cm to take advantage of the higher temperatures at the surface soil; later in the season, a depth of 6 cm is adequate. Germination occurs within 2–3 d if sufficient moisture is available. Because of insect hazards, spring planting is preferred over later planting in the season. Reduced tillage and the application of 5 t ha^{-1} straw to cover the seed bed significantly hastened seedling emergence and increased yields in irrigated silty soils with low water-infiltration rate. Sprinkler irrigation with or without mulch of these soils increased forage yield by approximately 25%, compared with flood or furrow irrigation (Tisdall and Adem, 1984).

Corn is seeded flat or on beds, usually in rows 75–90 cm apart. When drainage is a problem, 150-cm-wide beds may be used with two rows planted 55–57 cm apart on the bed and 90–95 cm between rows of adjacent beds (Kearney et al., 1980). Recommended plant density varies between 6 and 10 plants m^{-2}, depending on the hybrid and on growing conditions. Seeding rate should take into consideration a possible loss of 10–20% of the plants between sowing and harvest due to reduced germination, soil insects, birds, rodents, or adverse weather.

Double cropping is an accepted method for increasing silage production per unit area of land in Israel, when sufficient irrigation water is available. In this case an early-maturing forage-cereal for silage, usually wheat, precedes the corn crop. Autumn-seeded wheat is grown on approximately 2-m-wide beds. All wheel traffic at the time of wheat harvesting is confined to the tracks between the beds, thereby minimizing seedbed preparation for the following corn. Preplanting irrigation and rotary tilling to cope with excess stubble residue may sometimes be needed before the seeding of corn. Liquid fertilizer, especially N, is often applied concurrently with seeding.

Two successive seedings of trickle-irrigated forage corn have been tested in Israel with favorable results. Total dry-matter yields exceeded 40 Mg ha^{-1} and required ca. 600 mm irrigation water.

7.4 CROP WATER REQUIREMENTS AND IRRIGATION

7.4.1 Water-Production Functions

Evapotranspiration

The loss of water from the soil by evaporation during the early stages of crop growth is a major component of total ET. However, with the increase in leaf area and canopy closure daily water loss by transpiration soon dominates. Higher plant densities will increase more rapidly leaf area during early vegetative growth and result in increased water extraction. Water use varies according to the stage of plant development, with maximal consumption between tasselling and the dough stage. In the semi-arid environment of Clovis, New Mexico, potential daily water uses ranging from 7.5 to 9.8 mm d^{-1} were computed for corn (Abdul-Jabbar et al., 1983). Processes regulating water loss from corn leaves as related to soil water status were reviewed by Waldren (1983).

The relationship between DM yield and ET was found almost invariably to be linear (Hillel and Guron, 1973; Wenda and Hanks, (1981); Yanuka et al., 1982; Abdul-Jabbar et al., 1983). An example of this relationship is shown in Figure 7.4. A departure from linearity between yield and ET exists mostly for nutrient deficiencies (specifically of N) or excessive water (de Wit, 1958; Shimshi, 1966). In a trickle-fertigated experiment with forage corn in Israel, the yield was found to be a function of T and E_p:

$$DM = M(T/E_p)$$

where DM = dry matter yield; M = constant; T = daily transpiration; E_p = average daily pan evaporation). M had a value of approximately 290 kg DM ha^{-1} d^{-1} (Yanuka et al., 1982; Table 7.1).

Water Use Efficiency

A relatively wide range of ET-WUE values results from the differences in the environmental conditions under which experiments are carried out. The seasonal ET-WUE for well-watered corn was 1.2 kg DM m^{-3} at Bushland, Texas (Musick and Dusek, 1980) and 5 kg DM m^{-3} at Bet Dagan, Israel (Yanuka et al., 1982). In the latter case, the relatively high plant density and the avoidance of detrimental fluctuations in soil water and nutrient content were undoubtedly major factors contributing to this high WUE.

The use of fertilizer N coupled with short irrigation intervals increased the WUE of corn. More frequent irrigations apparently diminished losses

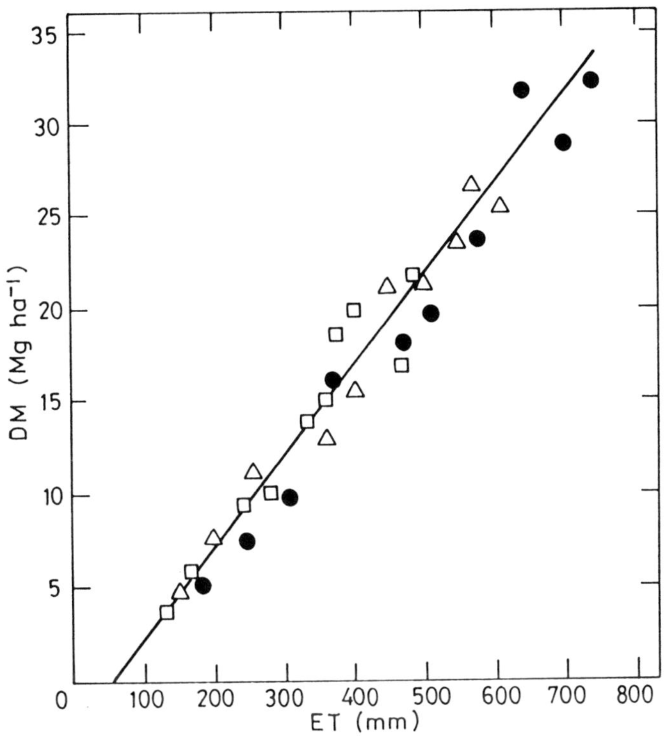

Fig. 7.4. Dry matter (DM) yield as a function of seasonal evapotranspiration (ET) in trickle-fertigated forage corn. Sampling at weekly intervals starting from the fifth week after seedling emergence. Total amounts of water applied: (●) 693 mm, (Δ) 459 mm, and (□) 230 mm. Bet Dagan, Israel. (Adapted from Yanuka et al., 1982).

TABLE 7.1

Dry matter yield (DM) of trickle-fertigated forage corn (cv. Newe Ya'ar 170), cumulative evapotranspiration (ET), mean Class A pan evaporation (E_p), and the calculated M value (= DM × E_p/ET) for different irrigation treatments, 106 d after seeding. Plant density 11 m^{-2}. Bet Dagan, Israel. (From Yanuka et al., 1982)

Water applied (mm)	DM (Mg ha^{-1})	ET (mm)	E_p (mm d^{-1})	M (kg ha^{-1} d^{-1})
693	32.4	730	6.6	293
459	25.8	613	6.6	278
235	22.1	489	6.6	298

due to denitrification and deep leaching (Russelle et al., 1981; Halitligil et al., 1984).

Crop Coefficient

Water consumption varies greatly throughout the growing season. The ratio of measured ET to water loss from a standard Class A evaporation pan can be used as a guide for determining the magnitude of water application. Doss et al. (1962) reported that the ratio of corn ET to pan evaporation varied from 0.38 at seedling emergence to 1.12 during the early dough stage, declining afterwards at grain maturity. Lomas et al. (1974) also reported that during the period of rapid plant development and with sufficient available soil water, the rate of water use may substantially exceed that measured in an evaporation pan. This may have occurred when atmospheric evaporative demands were particularly high. The common K_c values used in Israel for forage corn, according to growth stage, are: 0.4–0.6, seedling and early vegetative growth; 0.7–0.9, rapid vegetative growth until onset of tasselling; 1.0–1.2, pollination; 0.7–0.8, ear filling (Berkovitz, 1985).

7.4.2 Sensitivity to Water Stress

Root Growth

An extensive root system for the extraction of water and nutrients from the soil is a prerequisite for high forage yields. In irrigated well-drained loamy soils, roots have

been found to attain a lateral spread of up to 240 cm and a vertical penetration — if not limited by the soil profile — of at least 180 cm (Mantell, 1973). Lateral seminal roots begin to grow directly from the seed but slow down soon and are virtually non-existent when plants are approximately 20 cm high. A secondary (or nodal) root system starts to develop just below the soil surface, regardless of seeding depth, approximately one week after seedling emergence, but will not grow in dry soil. This secondary root system eventually becomes the major supplier of water and nutrients to the plant. At the end of the vegetative stage brace roots (also termed aerial nodal roots) grow from the nodes above the soil surface; they help to support the plant and scavenge the upper soil layers for water and nutrients during the reproductive stages (Kearney et al., 1980; Ritchie et al., 1992).

Mantell (1973) reported that until close to silking, the upper 120 cm of soil was thoroughly exploited by roots, but some water extraction occurred also below this depth. Approximately 55%, 33% and 12% of water was extracted by the roots of adequately-irrigated corn plants at the dough stage from the 0–60, 60–120 and 120–180 cm soil depths, respectively. Water deficits are generally reported to increase the absolute weight of the roots at the expense of the shoots (Rhoads and Bennett, 1990). Field observations suggest that corn is particularly susceptible to waterlogging damage at the first irrigation after emergence (Mason et al., 1987).

Growth Stages

The water stress response of corn is extensively reviewed by Rhoads and Bennett (1990). Generally, the sensitivity of crop growth stages to water deficits has been shown to decline in the following order: flowering and pollination > grain filling > vegetative growth (Musick and Dusek, 1980). Although this order of sensitivity to water stress was established for corn grown for grain, practical experience has shown that this is similar to silage corn.

Typical symptoms associated with water stress during the vegetative stage are:

(i) reduction in leaf expansion, occurring well before leaf photosynthesis is reduced (Boyer, 1970); and

(ii) reduction in internode lengths and plant heights, although the final number of leaves produced is generally unaffected by mild water deficits (Bennett and Hammond, 1983).

Water deficits during tasselling, silking and pollination are particularly detrimental to yield and may result in the delay of silking and inhibition of pollination. The primary effect of water stress during the ear-filling period is a reduction in current photosynthate supply, which is critical for optimum kernel fill. The duration, intensity and timing of stress obviously determine the magnitude of the reduction in yield.

7.4.3 Criteria for Irrigation Scheduling

Avoiding plant water stress dictates a high frequency of water application, whereas the amount of water to apply is determined by the depth of the root profile, and specific soil characteristics such as water-holding capacity.

From tasselling to silking, the depletion of available water should not be allowed to exceed 40–50% (Mantell, 1973; Musick et al., 1976). Overhead irrigation is often used until the crop is well established. The pre-seeding water application to a depth of ca. 120 cm is followed by small applications, the first after seeding and the second 4–5 d later, depending on soil type and weather conditions. Then trickle irrigation can be applied usually at 5-d intervals between irrigations, as required (Berkovits, 1985).

Limited Water Supplies

Most research indicates that corn, in contrast to sorghum, wheat or cotton, has little drought resistance, but uses water very efficiently. Thus, applying less water than is required to meet ET, in accord with growth stage, will appreciably reduce yield (Stewart et al., 1975; Musick and Dusek, 1980). Although the period around tasselling and silking is considered to be the most susceptible to water stress, even mild water deficits during other growth stages may diminish yields. The potential of using trickle-fertigation as a means of alleviating a deficient water supply for corn is not yet exploited sufficiently. Therefore, under conditions of limited available irrigation water, it appears essential to strive for maximum WUE and to avoid plant moisture stress in order to obtain maximum forage yield. This is, of course under the understanding that the other necessary agricultural inputs for high yield are also provided. Limited irrigation of corn is usually not feasible (Eck, 1986) and apparently leaves the grower no other option than to reduce the area for planting.

7.4.4 Sensitivity to Low Water Quality

Salinity

Maas (1987) classifies corn as moderately sensitive to salts in the irrigation water. Yield of corn is affected considerably by both soil water content and salinity of water application. Russo and Bakker (1987), examined the interaction between these two variables and found that the interaction should be described by a nonlinear concave expression, rather than the piecewise linear expression of Maas and Hoffman (1977). The response of corn yield to salinity in the water was examined experimentally and by modeling under conditions of high evaporative demands. Stover yields, irrigated with water of salinity of 1.8 dS m^{-1} dropped by 13 and 24% as a result of irrigation with saline water of 3.6 and 6.7 dS m^{-1}, respectively. The ear yield was more sensitive to salinity level than the stover (Russo and Bakker, 1987). Moreover, for the high salinity water of 6.7 dS m^{-1} a yield reduction occurred even when the amount of water applied was higher than maximum water uptake. This suggests that increasing the amount of irrigation water beyond the water use by the plants may compensate only in part for the adverse effects of the salinity of the irrigation water.

Sewage Water

Vaisman et al. (1986) report on the use of secondary municipal effluents (N: 40 mg l^{-1}; P: 35 mg l^{-1}: K: 50 mg l^{-1}) for the irrigation (at 2-d intervals) of forage corn on sand dunes in Israel. Total yields were ca. 17 Mg DM ha^{-1} and WUE was ca. 2.5 kg DM m^{-3} effluent. In experiments on a sandy loam soil, sewage effluent (N: 50 mg l^{-1}; P: 20 mg l^{-1}) was trickle-irrigated in addition to 60 kg N ha^{-1} applied as basic fertilizer on forage corn which yielded no less (ca. 23 Mg DM ha^{-1}) than fertilized plots receiving line water (Hefer and Itiel, 1990, not published).

7.5 MINERAL NUTRITION AND FERTILIZATION

Reviews by Arnon (1975), Olson and Sander (1988), and Rhoads and Bennett (1990) deal comprehensively with the nutritional requirements and fertilizer needs of grain corn, including irrigated corn. Many field studies established the relationships between the uptake of minerals — particularly N, P and K — and grain yield. Adequate levels of nutrients in leaves are shown in Table 7.2.

TABLE 7.2

Adequate levels of nutrients in the ear leaf of corn at silking. (From Barber and Olson, 1968)

Nutrient	g kg^{-1} (DW)
Nitrogen	28.0–33.0
Phosphorus	2.5–3.5
Potassium	17.5–22.5
Calcium	2.5–4.0
Magnesium	2.5–4.0
Sulfur	1.0–2.0
	mg kg^{-1} (DW)
Manganese	20–150
Iron	20–250
Boron	4–20
Copper	6–20
Zinc	20–70

stover. The improvement of stover quality, therefore, seems to be a key issue in the improvement of whole-crop digestibility. Unfortunately, however, low lignin content of the stem rind of the stem, increases the susceptibility to lodging (Bar-Tsur and Saadi, 1987).

7.7 HARVEST MANAGEMENT AND UTILIZATION

7.7.1 Silage

Corn is a widely grown silage crop under irrigation because of its palatability, high yield and high energy content. The optimum time for harvest is when kernels are at the late-dough stage (DM content 50–60%). The whole-crop DM content at this maturity level is 30–40%, depending on growing conditions. Harvesting and ensiling at this stage ensure optimal fermentation and minimal seepage loss, when procedures for filling the silo are followed strictly. Silage should be chopped from 0.5 to 1.5 cm long to facilitate packing, and crushed to break up corn kernels, thus reducing the passage of whole kernels through the animal's digestive system. Corn-silage management and other related subjects are reviewed by Hight and George (1983), McCullough and Bolsen (1984), Goodrich and Meiske (1985), Olson and Sander (1988) and Perry (1988).

The production of double-cropped corn silage is practised in the San Joaquin Valley of California. The cultural, pesticide and fertilizer inputs for the production of corn silage are similar to those for grain corn. Silage yields are estimated at 17 Mg DM ha^{-1} (Collar et al., 1990).

The feeding of corn silage to dairy and beef cattle as an energy supplement in production systems based on irrigated pasture, is reported from Australia (Moran and Trigg, 1985; Moran, 1986, 1989; Moran et al., 1986). In the so-called "Kyabram Dairy System", 40–50% of the diet is fed as corn silage complementary to grazed pasture at any stage of lactation, provided that the protein content and digestibility of the pasture are high. Pure legume pasture gives a higher marginal response to corn silage supplementation and enables more corn to be fed than do grass-dominant pastures. The growing of forage corn more than doubled the digestible energy per hectare and per unit of irrigation water than did grazed pasture (Pritchard et al., 1989). (See also Chapter 6, irrigated pasture.)

7.7.2 Green Chop

Freshly chopped corn is often harvested, hauled and fed to lactating cows when silo storage facilities are not available. This method permits harvesting at a near-ideal stage of maturity without storage cost and with minimal field losses, if suitable harvest equipment is used. However, the consistent supply of fresh corn to the stable or the feed center throughout the growing season, requires careful planning. It necessitates decisions on the timing of seedings, land preparation, and the size of the areas to be seeded sequentially in order to meet the daily forage demand. The green chop system, however, is not without its pitfalls and risks. Freshly harvested and hauled forage is prone to heating, spoilage and decrease of palatability if not fed immediately to the animals. Obviously, such losses should be prevented.

First of all, information is needed on the production potential of corn when seeded at different dates throughout the growing season. For example, a 3-year survey of ca. 60 field experiments in Israel showed that in seedings of forage corn made between early spring and mid-summer: (i) average DM yields decreased from 23 to 17 Mg ha^{-1}, (ii) average daily DM accumulation was more or less constant (210 kg ha^{-1}), and (iii) the number of days between seeding and harvest diminished from to 115 to 80 (Leshem, 1988). With this type of data the grower can map out a tentative seeding program to suit his farm conditions. Plant density is usually higher than when corn is grown for grain. Double-cropping of corn may increase the efficiency of water and land use.

Uniformity of forage quality, defined by maturity level at harvest and palatability of the cut forage during the growing season, is of prime importance in the feeding program of dairy cows. Thus, one of the crucial decisions to make is the interval of time between seedings (or when to start irrigation of a previously seeded field); or, alternatively, within how many days the harvest of a given field should be completed in order to maintain uniform forage quality, before starting to harvest the next field. Practical experience has taught that 7 d to a maximum of 10 d is the period during which only a relatively small change in forage quality occurs, although its DM content usually rises. Yields in excess of needs require an alternative means of conservation.

7.8 CONCLUSIONS

Forage corn is extensively cultivated under irrigation because of its high whole-crop yield and digestible energy content, water use efficiency, relative ease of cultivation, and short growth period. A high content of ears within corn forage facilitates ensiling, increases dry matter concentration and palatability, and reduces seepage losses in effluent. Recently developed hybrids have a greater tolerance to increased plant densities, and improved forage yields without a decrease in the ear fraction of the total plant mass.

Most investigations under nonstress conditions showed a near-linear relationship between evapotranspiration and whole-crop yield. Sensitivity to water deficits was found to be greater during flowering, pollination and kernel filling than during vegetative growth. High NPK levels in the soil throughout the growing season, the ability to maintain these levels, and an irrigation frequency which keeps soil tension low and nearly constant are important factors in achieving maximum yield and high water use efficiency. Corn is classified as moderately sensitive to salts in the irrigation water, and the ear yield is more sensitive to salinity level than the stover.

Differences in forage quality in terms of dry matter digestibility between hybrids are based mainly on differences between morphological plant fractions. The digestibility of kernels considerably outweighs that of other fractions with relatively high cell wall contents, such as leaf midrib and rind of the stem. The increase in highly digestible sugars in the pith of the stem often compensates for the decrease in ear-to-stover ratio as a result of high plant density. Improvement of stover quality appears to be a key issue in the improvement of whole-crop digestibility.

The whole-crop dry matter digestibility during the kernel-filling period is fairly uniform until the dough stage, after which it decreases. The time to harvest corn for silage therefore often depends more on considerations of appropriate dry matter content needed for adequate ensiling and yield of digestible matter, than on digestibility concentration *per se*. Double-cropping is an accepted method for increasing silage production per unit area of land, when sufficient irrigation water is available. In this case an early-maturing forage cereal, usually barley or wheat, precedes the corn crop.

The feeding of green-chopped corn permits harvesting at the optimal stage of maturity with minimal field and storage losses. However, this system requires careful planning of area size, sequential seeding and irrigation times, in order to meet the daily forage demand. Unforeseen surpluses, spoiled or lack of forage, are risk factors inherent to the system.

REFERENCES

Abdul-Jabbar, A.S., Sammis, J.W., Lugg, D.G., Kallsen, C.E. and Smeal, D. 1983. Water use by alfalfa, maize and barley as influenced by available soil water. Agric. Water Manage. 6: 351–363.

Akin, D.E. and Chesson, A. 1989. Lignification as the major factor limiting forage feeding value especially in warm conditions. 16th Int. Grassland Congr. (Nice, France) pp. 1753–1759.

Alberda, Th. 1984. Production and water use of several food and fodder crops under

irrigation in the desert area of southwestern Peru. Agric. Res. Rep. No. 928. Pudoc, Wageningen, The Netherlands.
Arnon, I. 1975. Mineral Nutrition of Maize. Int. Potash Inst., Bern, Switzerland, 452 pp.
Barber, S.A. and Olson, R.A. 1968. Fertilizer use in corn. pp. 163–188. In: L. B. Nelson et al. (Eds.) Changing Pattern of Fertilizer Use. SSSA, Madison, WI.
Barriere, Y. and Traineau, R. 1986. Characterization of silage maize: patterns of dry matter production, LAI evolution and feeding value in late and early genotypes. In: O. Dolstra and P. Miedema (Eds.) Breeding of Silage Maize 13th Congr. Eucarpia, Maize and Sorghum (Wageningen, The Netherlands) pp. 131–135.
Bar-Tsur, A. and Saadi, H. 1987. Prospects for breeding high quality forage corn. pp. 120–126. In: G. Ashbell and J. Palti (Eds.) Proc. Workshop on Replacement of Concentrates by Forage in the Diet of High-Producing Dairy Cows, Bet-Dagan, Israel.
Bar-Yosef, B. and Kafkafi, U. 1972. Rates of growth and nutrient uptake of irrigated corn as affected by N and P fertilization. Soil Sci. Soc. Am. Proc. 36: 931–936.
Beerepoot, L. 1981. The genetic variation in digestibility of forage maize. M.Sc. thesis, Agricultural University of Wageningen, The Netherlands. 43 pp. (in Dutch).
Bennett, J.M. and Hammond, L.C. 1983. Grain yields of several corn hybrids in response to water stresses impaired during vegetative growth stages. Proc. Soil Crop Sci. Soc. Florida. 42: 107–111.
Berkovitz, S. 1985. Maize Production. Agric. Ext. Service, Israel Min. Agric., Tel Aviv, Israel. 106 pp. (in Hebrew).
Boyer, J.S. 1970. Leaf enlargement and metabolic rates in corn, soybean and sunflower at various leaf water potentials. Plant Physiol. 46: 233–235.
Bunting, E.S. 1971. Plant density and yield of shoot dry material in maize in England. J. Agric. Sci. Camb. 77: 175–185.
Bunting, E.S. 1978. Agronomic and physiological factors affecting forage maize production. In: E.S. Bunting, B.F. Pain, R.H. Phipps, J.H. Wilkinson and R.E. Gunn (Eds.) Forage Maize Production and Utilization. Agric. Res. Council, London.
Cherney, J.H., Cherney, D.J.R., Akin, S.E. and Axtell, J.D. 1991. Potential of brown-midrib low-lignin mutants for improving forage quality. Adv. Agron. 46: 157–198.
Chevalier, P. and Schrader, L.E. 1977. Genotypic differences in nitrate absorption and partitioning of N among plant parts of maize. Crop Sci. 17: 897–901.
Collar, C., Higginbotham, G., Sutter, S., Vargas, R., Wright, S., Klonsky, K. and Livingston, P. 1990. Sample costs to produce double cropped corn silage in the San Joaquin Valley. Univ. of Calif. Crop Ext., Davis, CA. 8 pp.
Deinum, B. and Struik, P.C. 1986. Improving the nutritive value of forage maize. In: O. Dolstra and P. Miedema (Eds.) Breeding of Silage Maize. 13th Congr. Eucarpia, Maize and Sorghum (Wageningen, The Netherlands), pp. 77–90.
Dicke, F.F. and Guthrie, W.D. 1988. The most important corn insects. In: G.F. Sprague and J.W. Dudley (Eds.) Corn and Corn Improvement, ASA, CSSA and SSSA Monogr. 18: 767–867.
Doorenbos, J. and Kassam, A.H. 1979. Yield response to water. FAO Irrig. Drainage Pap. No. 33.
Doss, B.D., Bennett, O.L. and Ashley, D.A. 1962. Evapotranspiration by irrigated corn.

agronomic impacts of varied philosophies of soil testing. Agron. J. 74: 492–499.

Olson, R.A. and Sander, D.H. 1988. Corn production. In: G.F. Sprague and J.W. Dudley (Eds.) Corn and Corn Improvement. 3rd ed. ASA, CSSA and SSSA Monogr. 18: 639–686.

Perry, T.W. 1988. Corn as a livestock feed. In: G.F. Sprague and J.W. Dudley (Eds.) Corn and Corn Improvement. 3rd ed. ASA, CSSA and SSSA Monogr. 18, pp. 941–963.

Pritchard, K.E., Mason, W.K., Stockdale, C.R. and Moran, J.B. 1989. Integrating forage maize with pasture legumes for efficient dairy production in south eastern Australia. Proc. 16 Int. Grassland Congr. (Nice, France), pp. 1143–1144.

Rhoads, F.M. and Bennett, J.M. 1990. Corn. In: B.A. Stewart and D.R. Nielsen (Eds) Irrigation of Agricultural Crops. ASA, CSSA and SSSA Monogr. 30: 569–596.

Rhoads, F.M. and Stanley Jr, R.L. 1981. Fertilizer scheduling, yield, and nutrient uptake of irrigated corn. Agron. J. 73: 971–974.

Rhoads, F.M. and Stanley, R.L. Jr. 1984. Yield and nutrient utilization efficiency of irrigated corn. Agron. J. 76: 219–223.

Ritchie, S.W., Hanway, J..S. and Benson, G.O. 1992. How a corn plant develops. Iowa State Univ. Spec. Rep. 48. 21 pp.

Russelle, M.P., Deibert, E.J., Hauck, R.D., Stevanovich, M. and Olson, R.A. 1981. Effects of water and nitrogen management on yield and ^{15}N-depleted fertilizer use efficiency of irrigated corn. Soil Sci. Soc. Am. J. 45: 553–558.

Russo, D. and Bakker, D. 1987. Crop-water production functions for sweet corn and cotton irrigated with saline waters. Soil Sci. Soc. Am. J. 51: 1554–1562.

Schilfgaarde, J. van (Ed.) 1974. Drainage for Agriculture. ASA Monogr. 17. 700 pp.

Shaw, R.M. 1988. Climate requirements. In: G.F. Sprague and J.W. Dudley (Eds.) Corn and Corn Improvement, 3rd ed. ASA, CSSA and SSSA Monogr. 18: 609–638.

Shimshi, D. 1966. The effect of fertilization on some aspects of soil–water–plant relationships. Ph.D. thesis, The Hebrew University of Jerusalem, Rehovot, Israel. 145 pp. (in Hebrew).

Shurtleff, M.C. (Ed.) 1988. Compendium of Corn Diseases. American Phytopathology Soc., St. Paul, MI, 105 pp.

Smith, D.R. and White, D.G. 1988. Diseases of corn. In: G.F. Sprague and J.W. Dudley (Eds.) Corn and Corn Improvement. 3rd ed. ASA, CCSA and SSSA Monogr. 18: 688–766.

Sprague, G.F. and Dudley, J.W. (Eds.) 1988. Corn and Corn Improvement, 3rd ed. ASA, CCSA and SSSA Monogr. 18. 986 pp.

Stewart, J.I., Misra, R.D., Pruitt, W.O. and Hagan, R.M. 1975. Irrigating corn and grain sorghum with a deficient water supply. Trans. ASAE 18: 270–280.

Struik, P.C. 1982. Production pattern, chemical composition and digestibility of forage maize (*Zea mays* L.). Med. Vakgr. Landbouwgew. Graslandcult. 64. Wageningen, The Netherlands, 28 pp.

Struik, P.C. 1983. The effect of short and long shading applied during different stages of growth on development, productivity and quality of forage maize (*Zea mays*). Neth. J. Agric. Sci. 31: 101–124.

Struik, P.C., Deinum, B. and Hoefsloot, J.M.P. 1985. Effects of temperature during different stages of development on growth and digestibility of forage maize (*Zea mays* L.). Neth. J. Agric. Sci. 33: 405–420.

Tetio-Kagho, F. and Gardner, F.P. 1988a. Responses of maize to plant population density: I. Canopy development, light relationships, and vegetative growth. Agron. J. 80: 930–935.

Tetio-Kagho, F. and Gardner, F.P. 1988b. Responses of maize to plant population density: II. Reproductive development, yield and yield adjustments. Agron. J. 80: 935–940.

Tisdall, J.M. and Adem, H.H. 1984. The effect of reduced tillage and a mulch on seedling emergence and yield of maize (*Zea mays* L.) in an irrigated silty soil of south eastern Australia. pp. 169–174. In: W.A. Muirhead and E. Humphreys (Eds.) Root Zone Limitations to Crop Production on Clay Soils. Symp. Aust. Soil Science Soc. (Griffith, NSW, Australia).

Tollenaar, M. Daynard, T.B. and Hunter, R.B. 1979. Effect of temperature on rate of leaf appearance and flowering date in maize. Crop Sci. 19: 363–366.

Vaisman, I., Kipnis, T., Sharabani, H., Bielorai, H. and Feigin, A. 1986. The response of corn grown in sand dunes to irrigation with municipal effluents. Hassadeh 66: 2498–2501 (in Hebrew).

Versteeg, M.N. 1985. Factors influencing the productivity of irrigated crops in southern Peru in relation to prediction by simulation models. Ph.D. thesis. Agricultural University, Wageningen, The Netherlands.

Waldren, R.P. 1983. Corn. pp. 187–211. In: I.D. Teare and M.M. Peet (Eds.) Crop Water Relations. John Wiley & Sons, New York, NY.

Warrington, I.J. and Kanemasu, E.T. 1983. Corn growth response to temperature and photoperiod. 1. Seedling emergence, tassel initiation and anthesis. Agron. J. 75: 749–754.

Wenda, W.I. and Hanks, R.J. 1981. Corn yield and evapotranspiration under simulated drought conditions. Irrig. Sci. 2: 193–204.

Williams, W.A., Loomis, R.S., Duncan, W.G, Dovrat, A. and Nunez, F. 1968. Canopy architecture of various population densities and the growth and grain yield of corn. Crop Sci. 8: 303–308.

Wit, C.T. de 1958. Transpiration and crop yields. Versl. Landbouwk. Onderz. Med. 59. Wageningen, The Netherlands, 88 pp.

Yanuka, M., Leshem, Y. and Dovrat, A. 1982. Forage corn response to several trickle irrigation and fertilization regimes. Agron. J. 74: 736–740.

Chapter 8

BERSEEM CLOVER

8.1 INTRODUCTION

Berseem (bersim in Arabic) or Egyptian clover (*Trifolium alexandrinum* L.) is grown as a common forage crop in the Mediterranean basin, Iraq, Iran (40,000 ha, Vaez Zadeh (1989)) Pakistan and India. It is non-hardy and is grown mostly as a winter–spring annual under supplemental or full irrigation. In Egypt, for centuries berseem was considered indispensable as a half-year rotation crop with cotton, because of its high nitrogen-fixation capacity, (Richards, 1982). Alternately, it served as the principal source of fresh forage for draft animals and for dairy and beef cattle. At present, the area seeded to berseem is estimated at 700,000 ha (A.M. Rammah, personal communication). Especially for the smallholder farmer the cultivation of berseem as a major roughage supplier combines a number of attractive features, e.g.:

(i) it is superior in quality to conserved roughage when fed fresh;

(ii) field losses are kept to a minimum when the crop is periodically harvested and immediately fed to the animals;

(iii) a field can provide a constant supply of fresh feed for a period of 5–7 months, under appropriate climatic conditions;

(iv) it is easily harvested (often by scythe or sickle), although labor intensive; and

(v) it usually does not require nitrogen fertilizer.

Specific constraints in the cultivation of berseem are:

(i) the relatively low DM content of the forage during the autumn and winter months, and

(ii) the restriction in the use of mechanized harvesting implements on heavy-textured wet soils, which damage soil structure and lead to stand depletion.

Recently, in California and in the desert valleys of northern Mexico (Alvarez and Quiroga, 1992; Quiroga et al., 1993) there has been renewed interest in berseem as a forage producer and as a crop with nitrogen-fixing potential (Graves et al., 1987, 1989).

Berseem cultivation in northern Africa is reviewed by Gachet (1970), Gaillard et al. (1977) and Akremi and Graves (1981). The early introduction of berseem into California was reported by Kennedy and Mackie (1925).

The origin of this crop is not clear. All *T. alexandrinum* occurring in the wild are considered to be spontaneous forms originating from cultivation (Oppenheimer, 1959; Zohary and Heller, 1984; Dovrat, 1987). Cytogenetic studies of *Trifolium* spp. related to berseem were conducted by Putiyevsky and Katznelson (1970, 1972, 1973, 1974) and Putiyevsky et al. (1975).

8.2 GROWTH AND DEVELOPMENT

8.2.1 Morphology

Various features of the berseem plant are shown in Figure 8.1. The shoots are hollow and fleshy, can reach a height of 70 to 140 cm, and have an upright growth habit. Plants develop short taproots. Detailed botanical descriptions are provided by Zohary and Heller (1984) and Gillett (1985).

Cultivars of berseem are classified according to their branching behavior. Branching, in turn, influences the number of cuttings that can be made. The different branching behavior of types probably evolved as an adaptation to different amounts of water available for flooding in Egypt. 'Miscavi' and 'Kahdrawi' are basal (or crown) branching types which can yield four to seven cuttings (under irrigation) per growing season. The 'Saidi' type exhibits basal and apical, or stem branching and can be cut two to three times, whereas the 'Fahl' type is limited to apical branching and can be cut only once (Knight, 1985). Branching behavior within the above types is sometimes variable. A method to determine the trueness-to-type of different berseem seed lots was suggested by Nir (1961) when observing the regrowth pattern following the cutting of 45-day-old seedlings (Fig. 8.2). Also agronomically, these types differ among themselves in length of productive period, height, time of flowering, number of racemes per plant, and number of florets per raceme (Nir, 1961). Multiple-cut types are used mostly for green fodder, whereas 'Fahl' is used mainly for hay.

Fig. 8.1 (*Opposite*). The berseem clover plant. (A) Root system and shoots growing from the crown; (B) flowering branch; (C) enlarged flower. (From Zohary and Heller, 1984. Drawings by Dulic Amsler).

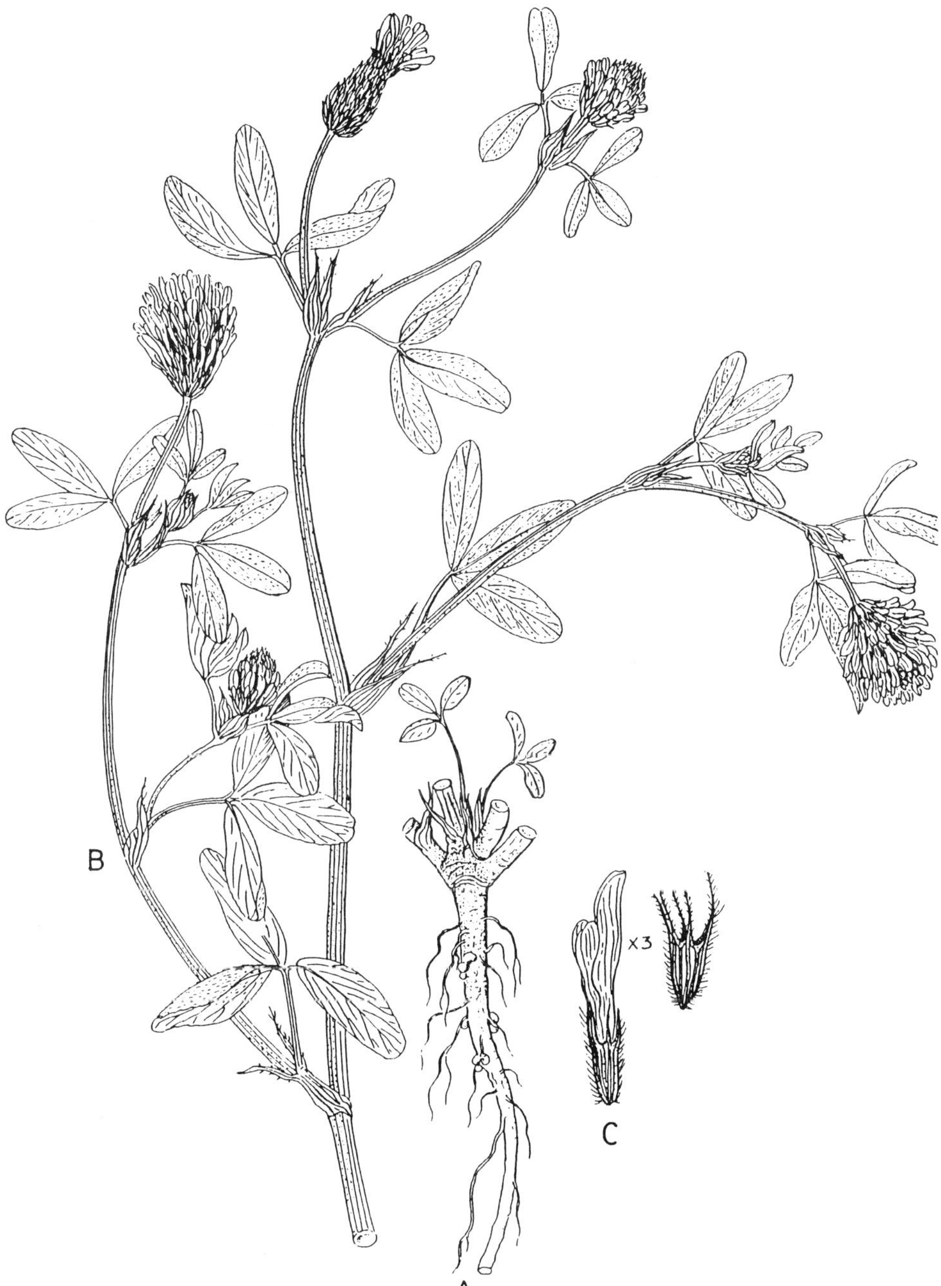
B
A
C
x3

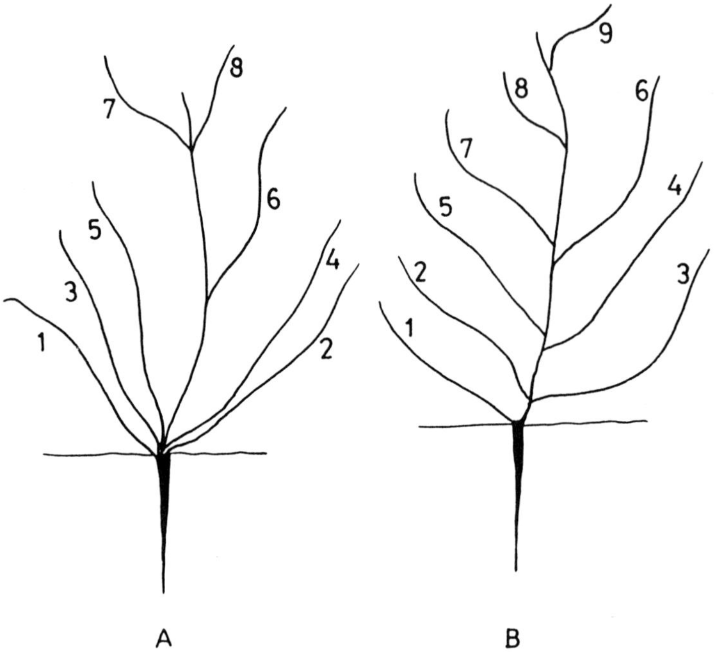

Fig. 8.2. Branching pattern of berseem. (A) Basal branching of 'Miscavi' multiple-cut type; (B) apical branching of 'Fahl' single-cut type. Numbers indicate sequence of branch appearance. (Adapted from Nir, 1961).

8.2.2 Canopy Structure

Shoot Initiation

Seedling development is similar to that of alfalfa which is described by Teuber and Brick (1988). No difference was found in seedling height between 'Fahl' and 'Miscavi' during the first 8 weeks of growth. 'Fahl' seedlings which have not been cut, continue to form secondary branches on the main stem (Fig. 8.3A), whereas 'Miscavi' seedlings develop new shoots from the crown and stubble buds after the first and successive cuttings (Fig. 8.3B). The greatest production of vegetative buds and DM yield was obtained when cv. Multicut was cut at approximately 7 cm above the soil surface (Marques, 1987). Close defoliation should be avoided. Also, frequent clipping sharply reduces DM production, and the number of plants and shoots (Hoveland and Andrews, 1966).

Vegetative Growth

'Miscavi' type berseem can be grown at temperatures from 0 to 35°C, but

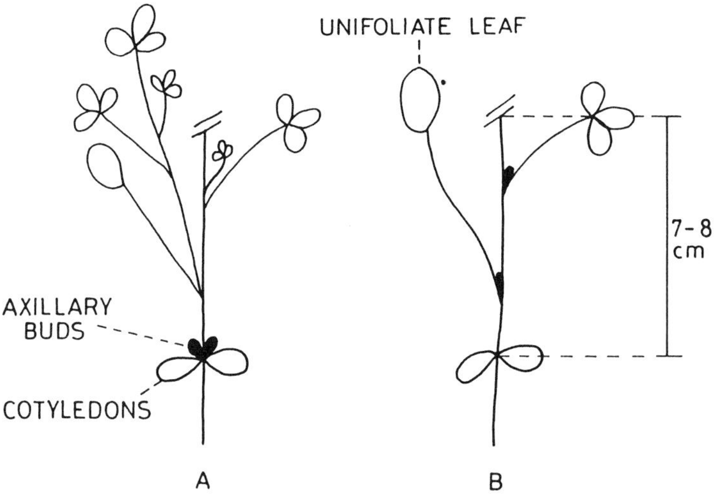

Fig. 8.3. Regrowth made during 10 d after 45-d-old berseem seedlings were cut at 7–8 cm height above ground level. (A) 'Miscavi', multiple-cut type. Further development of the primary stem arising from the axillary bud of the unifoliate leaf, and bud initiation at the cotyledonary nodes. (B) 'Fahl', single-cut type. Initiation of axilliary buds of the unifoliate leaf and of the secondary stem only. (Adapted from Nir, 1961).

the optimal range is 12–25°C. Winter-hardy cultivars like 'Multicut' and 'Bigbee' have withstood, without damage, temperatures as low as –5°C (Graves et al., 1987). Berseem is usually harvested when plants are 40–50 cm high. At this plant height the maximum rate of DM production and the maximum leaf-to-stem ratio are reached, and crown buds, similar to those of alfalfa, have already started to develop. At this stage, renewed basal growth contributes little to the yield, but is essential for regrowth after cutting. Delays in harvesting reduce total forage yields. A typical example of average daily growth rates during five successive growth cycles is presented in Figure 8.4, which shows that the daily yield increments increased from autumn to spring from approximately 40 to 65 kg dm ha^{-1}. Concurrently, the DM concentration of the forage increased from a low of 100 g kg DM ha^{-1} during the autumn and winter, to almost double in the spring, towards the end of the growing season. Growth remains vegetative until spring, when flowering occurs.

Reproductive Growth

Floral induction of autumn-seeded 'Miscavi' takes place in spring with the rise in temperature and increase of daylength to over 12 h. Flowering starts usually in late spring. 'Fahl' has a much less pronounced thermo-

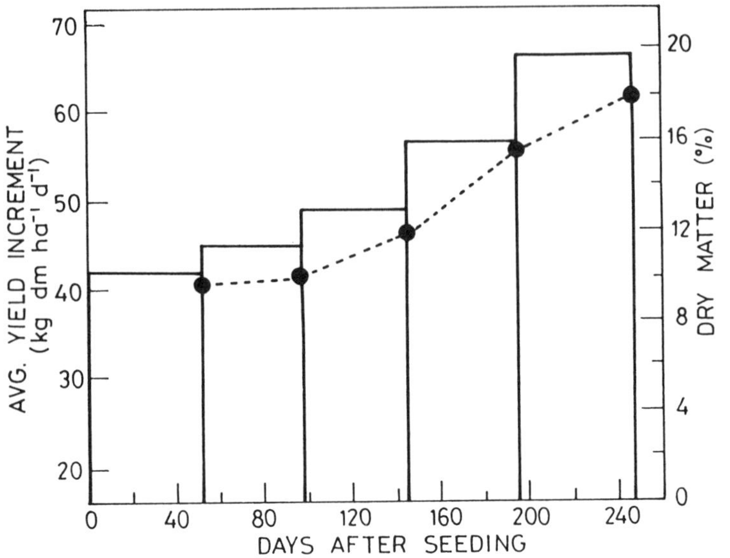

Fig. 8.4. Average daily dry matter increment and dry matter concentration of berseem cv. 'Carmel' grown under supplemental irrigation. Seeding date, 20 September 1975. Total dry matter yield, 12.9 Mg ha^{-1}. Beer Toviyya, Israel.

photoperiodic response and flowering usually starts 3 months after the sowing date.

8.3 ESTABLISHMENT

Soil Requirements

Berseem is cultivated on a wide variety of soils, including calcareous and heavy soils. It is quite tolerant of salinity, except in the early stages of growth. A pH of 6–7 is needed for the proper functioning of N_2-fixing *Rhizobium* bacteria.

Tillage and Seedbed Preparation

No special tillage is required for berseem, but a friable seedbed is desirable. The land is plowed to a depth of 15–20 cm, disked and leveled. Recommended seed rates are 40–50 kg ha^{-1} (Postiglione, 1966; Farid et al., 1968), and the optimal seeding depth is 1–2 cm, with a maximum of 5 cm. The seeds germinate very rapidly under favorable conditions and the seedlings are relatively strong (Arnon, 1972).

Inoculation

In soil known to lack efficient rhizobia for nodulation, the seeds should be inoculated with *Rhizobium trifolii* specific to berseem clover.

Date and Methods of Sowing

In the Mediterranean region the optimal sowing time for multiple-cut cultivars is from late-September to mid-October. Earlier sowing often prevents proper seedbed preparation and increases the incidence of caterpillar and bird damage; early sowings also often suffer from excessive heat and require large amounts of irrigation water. In later sowings the initial growth, number of cuts and total forage yields are smaller than from sowings at optimal dates (Arnon, 1956; Restuccia and Longo, 1968; Gaillard et al., 1977). Seed is either broadcast or drilled (in rows spaced 15–20 cm apart), before or after irrigation. In India, berseem is sometimes sown after the harvest of the paddy (rice) crop. In that case, seed is broadcast when the soil surface is still covered with a thin layer of water (Tiwana, 1980).

8.4 CROP WATER REQUIREMENTS AND IRRIGATION

8.4.1 Water-Production Functions

Evapotranspiration

Only few measurements of ET of berseem are available. Gaillard et al. (1977), using lysimeters, found that DM yield was a quadratic function of ET during the spring months in Algeria. In their study, the calculated *water use efficiency* (WUE) ranged from 2.3 to 1.6 kg DM m^{-3} from the beginning of the spring until the last cutting 3 months later. Similar results are reported from northern Mexico (Alvarez and Quiroga, 1992). Berseem WUE was also calculated from DM yields obtained from field experiments at Davis, California (Williams et al., 1991; Fig. 8.5). The estimated WUE values ranged from 1.8 to 2.6 kg DM m^{-3} water used in the autumn–winter harvests and dropped to an average of 1.3 kg DM m^{-3} for the spring harvests, as evaporative demand increased. These values resemble those from alfalfa, according to published studies (see Table 5.4). However, unlike alfalfa, berseem growing in the Mediterranean basin, has the advantage of being produced primarily during the rainy season before ET reaches a summer peak.

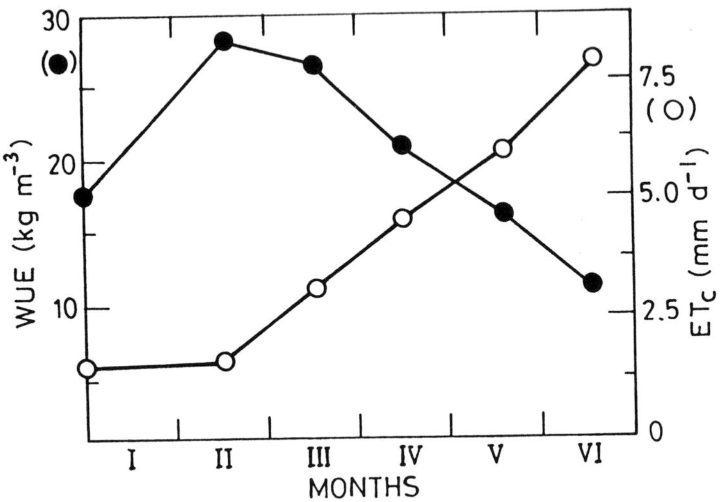

Fig. 8.5. Estimated water use efficiency (WUE) of berseem cv. "Multicut' and crop evapotranspiration (ET), based on daily means for six growing seasons at Davis, California. (Adapted from Williams et al., 1991).

8.4.2 Sensitivity to Water Stress

Root Growth

Few reliable data on rooting depth are available. The major part of the root system is found in the top 40–50 cm soil layer, but roots can reach a depth of 90 cm or more (Kurtz, 1974). Depth of rooting is barely affected by the number of harvests (Roy Sharma and Singh, 1969).

Because of berseem's relatively shallow root system, frequent irrigations are required in order to restore depleted available soil moisture. However, too frequent or heavy irrigations, particularly in autumn, may contribute to waterlogging during the winter months in winter-rainfall areas. Waterlogging results in a high incidence of root rots, a depleted stand and reduced yields throughout the growing period. In winter-rainfall areas, wetting of the entire potential root zone before sowing is therefore not recommended (Arnon, 1972).

Moisture Stress

Berseem is very sensitive to the soil moisture regime: yields and the length of the growing period are markedly affected by irrigation. Water stress in the spring can cause early flowering, which is associated with decreased vegetative growth resulting in low forage and seed yields, even when adequate irrigation is restored. Hot, dry winds and the absence of available soil water cause shriveling and desiccation of the leaflets.

8.4.3 CRITERIA FOR IRRIGATION SCHEDULING

At the pre-sowing irrigation, it is not recommended to wet the soil below a depth of 60 cm. After sowing, light irrigations are given to ensure proper germination and seedling emergence. This is necessary particularly in soils which have the tendency upon drying to form a surface crust layer, which prevents seedling emergence. After establishment, successive irrigations are applied at about fortnightly intervals until the start of the rainy season.

With the onset of the dry season, irrigations are resumed every 14–7 d with gradually increasing amounts of water according to the increase in the evaporative demands, until the last harvest. The amount of water given will depend on the required wetting depth and the water-holding capacity of the soil.

8.4.4 Sensitivity to Low Water Quality

Salinity

Berseem is considered to be moderately salt resistant (Winter and Lauchli, 1982). Gachet (1970) and Noble and West (1987) quote a threshold EC level of approximately 3 dS m^{-1} for initial yield decline. The first cutting usually has a lower threshold level than subsequent cuttings. In newly-reclaimed saline areas in Egypt, berseem is among the first crops to be planted, generally following a rice crop (Kaddah, 1962). Subsurface drainage, gypsum application and deep plowing of fields in the Nile Delta, significantly improved the berseem yields in the regular crop rotation. Soil salinity to a depth of 1.5 m was reduced from an average 5.3 to 2.2 dS m^{-1} (El-Mowelhi et al., 1988). Winter and Lauchli (1982) and Winter (1982) grew berseem plants in solution culture as well as in the field and found that chlorine ions (Cl^-) accumulated in the roots with respect to the shoot, indicating their controlled transport, in particular to young leaves.

The availability of genetic variation for salt tolerance seems to offer possibilities for breeding purposes. Ashraf et al. (1987) reported on the selection of berseem lines which produced significantly higher DM yields than unselected lines. Tolerant lines generally contained less sodium ions (Na^+) ions and potassium ions (K^+) but more Cl^- in their shoots than normal lines. They concluded that significant responses to recurrent selection for seedling NaCl tolerance may be expected in this species. These

Technology. Am. Soc. Agron. Monogr. 25: 26–27.

Goldman, A., Yulzari, A., Genisi, A., Guggenheim, Y. and Lev, B. 1985. Digestibility and protein in roughages in Israel. Min. Agric. Ext. Serv. Tel Aviv, Israel. 35 pp. (in Hebrew).

Graves, W.L., Williams, W.A., Wegrzyn, V.A., and Calderon M.D. 1989. Registration of 'Multicut' berseem clover. Crop Sci. 29: 235–236.

Graves, W.L., Williams, W.A., Wegrzyn, V.A., Calderon M.D., George, M.R. and Sullins, J.L. 1987. Berseem clover is getting a second chance. Calif. Agric. 41 (9/10): 15–18.

Guessous, F. 1982. Age, date of cutting, and temperature as factors affecting chemical composition of berseem (*Trifolium alexandrinum* L.). 14th Int. Grassl. Congr. (Lexington, KY), pp. 475–478.

Hoveland, C.S. and Andrews, O.N. 1962. Growth of berseem clover as influenced by clipping treatment. Crop Sci. 2: 368.

Kaddah, M. 1962. Tolerance of berseem clover to salt. Agron. J. 54: 421–425.

Kennedy, P.B. and Mackie, W.W. 1925. Berseem or Egyptian clover (*Trifolium alexandrinum*). Univ. Calif. Bull. 389. 31 pp.

Knight, W.E. 1985. Miscellaneous annual clovers: berseem clover. In: N.L. Taylor (Ed.) Clover Science and Technology, Am. Soc. Agron. Monogr. 25: 547–551.

Kurtz, H. 1974. Irrigation of forage crops. Min. Agric. Ext. Serv., Tel Aviv, Israel. 13 pp. (in Hebrew).

Labban, F.M., Aboul-Naga, A. and Radwan, A. 1971. A study of the incidence of bloat from grazing Egyptian clover in Merino "Fleisch Schaf", Ossimi, Barky and crossbred flocks. Agric. Res. Rev. Cairo 49 (4): 61–68.

Latif, L.I.A. and Sinha, M.N. 1983. Studies on yield and nutrient uptake by berseem as influenced by NPK fertilization. J. Res. Haryana Univ. (Egypt) 13: 230–237.

Lev, B. 1982. Feed composition and nutritive value of feedstuffs for dairy cattle, other ruminants and horses. Min. Agric. Ext. Serv. Tel Aviv, Israel. 23 pp. (in Hebrew).

Lowe, K.F. and Bowdler, T.M. 1982 Note on the effect of sowing date, defoliation management and closure date on the seed production and forage yield of berseem clover (*Trifolium alexandrinum*). Trop. Grassl. 16: 40–43.

Marques, H. 1987. Effect of production of vegetative buds on forage yield by three berseem clover varieties. M.Sc. thesis, Univ. of California, Davis, CA. 31 pp.

Nir, J. 1961. Morphological and physiological characteristics of Egyptian clover varieties. M.Sc. thesis, Hebrew University of Jerusalem, Rehovot, Israel. 65 pp. (in Hebrew).

Noble, C.L. and West, S. W. 1987. Maximizing clover growth in the presence of salinity. In: J.L. Wheeler, C.J. Pearson and G.E. Robards (Eds.) Temperate Pastures: Their Production, Use and Management. CSIRO, Melbourne, Australia. pp. 123–125.

Oppenheimer, H.R. 1959. The origin of Egyptian clover with critical revision of some closely related species. Bull. Res. Counc. Isr. Sec. D7: 202–221.

Postiglione, L. 1966. Un biennio di esperianze sul trifolio alesandrino. Ann. Fac. Sci. Agrar. Univ. Napoli, Italia 4: 189–217.

Putiyevsky, E. and Katznelson, J. 1970. Chromosome number and genetic system in several *Trifolium* species related to *T. alexandrinum*. Chromosoma 30: 476–482.

Putiyevsky, E. and Katznelson, J. 1972. Cytology and crossability of several Mediterranean *Trifolium* species. Israel J. Bot. 21: 179–181.

Putiyevsky, E. and Katznelson, J. 1973. Cytogenetic studies in *Trifolium* spp. related to berseem I. Intra- and interspecific hybrid seed formation. Theor. Appl. Genet. 43: 351–358.

Putiyevsky, E. and Katznelson, J. 1974. Cytogenetic studies in *Trifolium* sp. related to berseem III. The relationships between the *T. scutatum*, *T. plebeium* and the *echinata* group. Theor. Appl. Genet. 44: 184–190.

Putiyevsky, E., Katznelson, J. and Zohary, D. 1975. Cytogenetic studies of *Trifolium* spp. related to berseem IV. The relationships in the *alexandrinum* and *vavilovi* crossability groups, and the origin of the cultivated berseem. Theor. Appl. Genet. 45: 355–362.

Quiroga, H.M., Cueto, J.A. and Jimenez, J. 1993. Berseem clover a new forage crop for the dairy farms in Mexico. Proc. 17th Int. Grassl. Congr. (Australia–New Zealand) (in press).

Rafey, A., Shahani, M.N. and Khan, A.R. 1978. Effect of different levels of phosphatic and potassic fertilizers on the quality of berseem forage. Fert. Techn. 15: 73–77.

Rammah, A.M. and Radwan, M.S. 1977. The influence of seeding rate and cutting management on yield and botanical composition of a berseem-grass mixture. J. Agric. Crop Sci. 145: 103–111.

Restuccia, G. and Longo, G. 1968. Epocadi semina e frequenza dei tagli nella tecinca di coltivazione del trifolio alesandrino. Tec. Agrar. 20: 581–598.

Richards, A. 1982. Egypt's Agricultural Development, 1800–1980. Technical and Social Change. Westview Press, Boulder, CO. 296 pp.

Rotili, P. and Gnocchi, G. 1989. The effect of selfing on *Trifolium alexandrinum* measured in conditions of interference among the plants. 16th Int. Grassl. Congr. (Nice, France), pp. 307–308.

Roy Sharma, P.R. and Singh, A. 1969. Studies on the frequency of clipping of berseem on the yield of forage root growth and nitrogen content of roots. Indian J. Agric. Sci. 39: 886–892.

Shafashak, S.E., El-Debaby, A.S., Salem, M.S. and Gelilah, G.Y. 1982. Crop rotation studies in Egypt: II. Growth and yield of Egyptian clover as affected by crop rotation and fertilizer residues. Ain Shams Univ. Fac. Agr. Res. Bull. 2089, 20 pp.

Sims, J.R., Solum, D.J., Westcott, M.P., Jackson, G.D., Kushnak, G.D., Wichman, D.M., Welty, L.E., Berg, R.K., Eckhoff, J.L., Stallknecht, G.F. and Gilberthen, K.M. 1991. Yield and bloat hazard of berseem clover and other forage legumes in Montana. Montana AgResearch Univ. Agr. Exp. St. pp. 4–10.

Singh, A.P. and Gupta, S.C. 1971. Studies on the nutrient yield and possible toxic and other physiologically active constituents in forage crops. Ann. Rep. 1971 Indian Grassl. Fodd. Res. Inst., Jahusi.

Sinha, N.C. and Singh, R.P. 1986. Physiological basis of irrigation scheduling for seed production in Egyptian clover syn. berseem crop. J. Agric. Crop Sci. 156: 246–252.

Teuber, L.R. and Brick, M.A. 1988. Morphology and anatomy. In: A.A. Hanson, D.K. Barnes and R.R. Hill (Eds.) Alfalfa and Alfalfa Improvement, Am. Soc. Agron. Monogr. 29: 125–162.

Tisserand, J.L. and Alibes, Y. 1989. Feeds of the Mediterranean area. pp. 305–323. In: R. Jarrige (Ed.) Ruminant Nutrition: Recommended Allowances and Feed Tables. INRA, Paris.

Tiwana, M.S. 1980. Berseem. pp. 30–37. In: Fodder Crops. Punjab Agric. Univ., Ludhiana, India.

Tiwana, M.S., Puri, K.P. and Thind, I.S. 1989. Competitive effects on production of *Lolium perenne* and *Trifolium alexandrinum* in mixture and monoculture. pp. 623–624. 16th Int. Grassl. Congr. (Nice, France).

Vaez Zadeh, A. 1989. Determination de l'effet du phosphore et des intervalles d'irrigation sur la production de graines de trefle bersim en Iran. pp. 657–658. 16th Int. Grassl. Congr. (Nice, France).

Williams, W.A., Graves, W.L. and Cassman, K.G. 1990. Nitrogen fixation by irrigated berseem clover versus soil nitrogen supply. J. Agric. Crop Sci. 3: 202–207.

Williams, W.A., Graves, W.L., Cassman, K.G., Miller, P.R. and Thomsen, G.D. 1991. Water-efficient clover fixes soil nitrogen, provides winter forage crop. Calif. Agric. 45 (4): 30–32.

Winter, E. 1982. Salt tolerance of *Trifolium alexandrinum* L. II. Ion balance in relation to its salt tolerance. Aust. J. Plant Physiol. 9: 227–237.

Winter, E. and Lauchli, A. 1982. Salt tolerance of *Trifolium alexandrinum* L. I. Comparison of the salt response of *T. alexandrinum* and *T. pratense* L. Aust. J. Plant Physiol. 9: 221–226.

Zohary, M. and Heller, D. 1984. The Genus *Trifolium*. Israel Academy of the Sciences and Humanities, Jerusalem, 606 pp.

Chapter 9

ANNUAL RYEGRASS

9.1 INTRODUCTION

The genus *Lolium* dominates the natural grasslands of temperate Europe, temperate Asia and North Africa and is and introduced crop in temperate and subtropical areas throughout the world. The perennial form *L. perenne* is a common constituent of short to long-grazing leys (also irrigated) and permanent pastures. The annual or biennial form, *L. multiflorum*, originates in western Europe, North Africa and southwest Asia and is adapted to mild, humid, temperate or subtropical climates and to moist, well-drained, fertile soil (Whyte et al., 1959). Two cultivated forms of annual or common ryegrass are recognized:

(i) Italian ryegrass (*L. italicum* or *L. multiflorum* var. *italicum*), which is short-lived and usually an annual, and

(ii) Westerwold ryegrass (*L. multiflorum* var. *westerwoldicum*), which is a rapid-growing, strictly annual, broadleaved strain.

Morphologically, these two forms are very similar. Many cultivars, including tetraploid ones, were developed from each of these forms, differing in leafiness, tillering capacity, growth habit, cold and disease resistance, and regrowth potential after defoliation.

Irrigated annual ryegrass (AR) is capable of producing very high forage yields with excellent nutritional value, and offers ample diversity in utilization options. AR is cultivated, or under trial, mainly in regions with mild winters, e.g. Algeria (Ahmin et al., 1975), Morocco (Ameziane, 1979; Mejjati-Alami et al., 1989), Israel (Dovrat and Lachover, 1958; Leshem et al., 1986), Yemen Arab Republic (Farnworth and Said, 1983), Saudi Arabia (Bartholomew and Williams, 1981), USA (Hagemann et al., 1984; Graves et al., 1987), Italy (Rotili et al. 1985), France (Reyne et al., 1977) and other countries. In these countries it is generally cut, hauled and fed fresh during the winter and made into hay, haylage or silage during the spring. Often it is a complementary feed to berseem clover.

Annual ryegrass is mostly grazed, either seeded alone or in mixture with legumes, in Australia (Robbins et al., 1980; Lowe et al., 1983; Lowe and Hamilton, 1985; Robbins and Bushel, 1987), South Africa (Goodenough et al., 1984; Smith et al., 1988), northern Mexico (12,000 ha, Graves and Munoz, 1987) and Spain (Rosello, 1971; Garcia-Criado et al., 1984). Special efforts should be made to resolve the problems related to:

(i) high costs for the purchase of N fertilizers,

(ii) development of technologies for the curing of hay during the rainy season,

(iii) sensitivity to wheel traffic of heavy harvesting equipment, and

(iv) susceptibility to leaf diseases.

The production potential of AR in northern Morocco and in Israel has been discussed by Ameziane (1979) and Leshem et al. (1989), respectively.

9.2 GROWTH AND DEVELOPMENT

Morphological and physiological aspects of temperate grasses, including AR were reviewed by Robson et al. (1988).

9.2.1 Canopy Structure

Juvenile

The optimum temperature and moisture requirements for germination and radicle growth of AR are similar to those of other temperate grasses. The rate of seedling emergence increases with temperature (Skjelvag, 1986). Under favourable growing conditions, seedlings soon start tillering until a LAI of approximately 3 is reached, corresponding to approximately 6,000 tillers per m^2, independent of sowing density. This LAI value coincides with the almost complete extinction of light at the level of the tiller buds close to the soil surface (Simon, 1981; Simon and Lemaire, 1987). Thus, a full ground cover can be expected to develop even after a low-density seeding, but more time is required to attain it. Relatively high seeding rates, however, are recommended to ensure rapid leaf cover of the soil surface, which will diminish evaporation losses and increase the light interception capability of the canopy. LAI values continue to increase after a full ground cover is reached.

Vegetative Growth

During the period of vegetative growth, throughout the autumn, winter and spring, plants are usually exposed to a rather large range of temperatures. The optimum temperature for most aspects of growth is between 20 and 25°C, as for most temperate grasses (Robson et al., 1988). Under desert conditions with relatively high insolation, Sagi (1992) found a significant linear relationship between DM accumulation of field-grown AR and growing degree days (GDD) during the autumn to spring period of growth. Thus, temperature seems to be an overriding factor affecting growth rate when other growth factors are optimal. Ameziane (1979), Ahmin et al. (1975) and Leshem (1976) report average daily growth rates of 40–80 kg DM ha^{-1} d^{-1} during the linear period of growth in the spring, with an almost twofold increase during the reproductive stage of growth. As a result of these differences between growth rates, the yield during vegetative growth was about half of the total DM of the entire production period.

Reproductive Growth

Annual ryegrass has a long-day flowering response but no obligatory cold requirement for flower initiation. The shoot apex develops flower primordia rather early in the growing season but spikes usually do not emerge before the beginning of spring (Aitken, 1985). Flowering is associated with a decrease in forage quality because of the presence of less-digestible stems. Flowering is usually also the period of severe rust infection of leaves and stems.

Seed production is a specialized activity and not recommended under irrigated conditions. Seed yields are usually low, because of the seed shattering and the incidence of leaf diseases.

9.3 ESTABLISHMENT

Soil Requirements and Seeding

Annual ryegrass has a wide range of soil adaptability, but for optimum forage production it does best on permeable soils with medium to high fertility. The seeding of AR on soils prone to waterlogging should be avoided because of the crop's sensitivity to inundation even for short periods (Ameziane, 1979; Goren and Kessler, 1980). AR has a reputation for easy and quick establishment, if seeded at a shallow depth on a well-prepared and firm seedbed. The recommended seeding rate is 40–60 kg ha^{-1}, but tetraploid cultivars are seeded at a somewhat higher rate because of the greater weight of the seeds.

Soil wetting to a depth of 20–30 cm is required after seeding, followed by the application of relatively small amounts of irrigation water at 2–3 d intervals until full establishment is achieved.

soil water deficits, depending on soil characteristics. Under conditions of high evaporative demand irrigation may be required once in 7–10 d. Water to flood-irrigated plots on Imperial silty clay soil was applied when the soil matric potential at a 25-cm depth decreased below –0.05 MPa, as indicated by tensiometers. Under desert conditions of the northern Negev in Israel the total water requirement is estimated at 500–600 mm (Sagi, 1992).

Limited Water Supplies

Among the options available to the grower when the amount of irrigation water is limited, are: (i) restricting the size of the area seeded; (ii) exploiting the autumn to early spring period for forage production and withdrawal of water at the onset of flowering, if forage quality is of overriding importance; (iii) deferring seeding to late autumn, which also facilitates hay- and silage-making operations during the spring. It should be noted, however, that none of the above-mentioned options has been critically studied in terms of crop WUE.

9.4.4 Sensitivity to Low Water Quality

Saline Water

Maas (1985) rated the closely related perennial ryegrass as moderately tolerant, with a soil salinity threshold of 5.6 dS m^{-1}. Field experiments in Israel confirmed these findings with AR. Leshem et al. (1989) and Sagi (1992) studied the yield response of AR to saline water concentrations in line-source irrigation experiments on loessial soil during 2 years in the northern Negev of Israel. Dry matter yields did not decrease with saline water concentration below 5.5 dS m^{-1}, but above this concentration yield decreased, in comparison with a non-saline-water-irrigated crop. *In vitro* DMD of the forage was not affected following irrigation with water containing 6.2 dS m^{-1}.

Sewage Water

Sewage effluent, with an average total N of 51 g l^{-1} was used for sprinkler irrigation of AR in the northern Negev and supplied sufficient nutrients for optimal forage yields (Leshem et al., 1986).

9.5 HARVEST MANAGEMENT

Cutting Intervals

The number of harvests will depend on the date of sowing, utilization and quality requirements. For green chop or fresh feeding during the

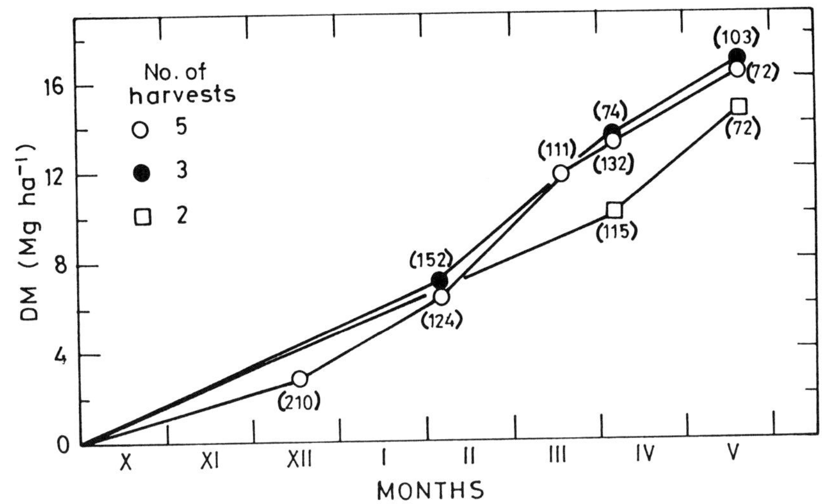

Fig. 9.2. Effect of number of harvests on dry matter yield and crude protein concentration (in parentheses, g kg^{-1}) of irrigated annual ryegrass under desert conditions at Ramat Hanegev, Israel. Total crude protein yields were 2.3, 1.8 and 1.5 Mg ha^{-1} from 5, 3 and 2 harvests, respectively. Date of seeding: October 20. Date of spike emergence: April 7. (Adapted from Sagi, 1992).

period of vegetative growth, harvesting is usually carried out as soon as fresh yields reach 3–4 Mg ha^{-1}. Cutting will have to be delayed if rain incidence limits harvest and transport procedures. The number of harvests was found to have little effect on total DM yields in Israel, except when harvesting started at the beginning of reproductive growth, (Sagi, 1992, Fig. 9.2). Thus, in hay and silage operations, two or three harvests will suffice to obtain maximum DM yields, save harvesting and transport costs, and diminish wheel traffic damage. However, forage quality will be less than with four or five harvests. Forage quality gradually decreases as the season progresses and more so when intervals between harvests are extended. Details of cultivation and representative yields from the Mediterranean region are shown in Table 9.1.

9.6 MINERAL NUTRITION AND FERTILIZER USE

Maximum DM and protein yields require copious amounts of mineral nutrients in general and of N in particular.

Nitrogen

Nitrogen is a dominant element in the yield formation of irrigated AR if

TABLE 9.1

Cultivation details and yields of irrigated annual ryegrass in the Mediterranean region. (Adapted from Ameziane (1979) and Sagi (1992))

Location	Seeding date	Spike emergence	Date 1st harvest	Total no. harvests	Total yield ($Mg\ ha^{-1}$)	Reference
Saragossa, Spain	Sept. 7	April 22	Dec. 9	5	11–18	Rosello (1971)
Montpellier, France	Sept. 3–13	April 28	End Feb.	3	14	Reyne et al. (1977)
Mitidja, Algeria	Sept. 26	May 5	Dec. 12	3–6	14	Ahmin et al. (1975)
Gharb, Morocco	Sept. 22	May 1	Dec. 7	3–4	14–18	Amiziani (1979)
Ramat Hanegev, Israel	Oct. 2	April 7	Dec. 15	4–5	15–18	Sagi (1992)

the availability of other elements is not limited. For instance, increments of 225 kg fertilizer N ha^{-1} significantly increased DM yields up to a total of 675 kg fertilizer N, and higher rates maintained yields longer. On the other hand, in unfertilized plots, the DM yield of AR for the entire season was negligible when preceded by a summer planting of Sudan grass which had almost completely depleted soil N reserves (Fig. 9.3). Similar yield responses to fertilizer N are reported from Israel (Leshem, 1976). Exceptionally high net uptake of N by AR was found also in experiments carried out on reclaimed desert soil in Saudi Arabia and no variance was found in the response to N fertilizers differing in formulation (Bartholomew and Williams, 1981).

The greedy mode of N uptake by AR plants was demonstrated also in experiments conducted on deep, well-drained, fertile sandy-loam soils in the presence of residual soil N in California. The first (and sometimes the second) harvest of AR exceeded that of berseem, but the yields of the successive harvests of AR dropped drastically in the absence of fertilizer N (Graves et al., 1987). This characteristic of a high potential uptake for soil N by AR was employed to estimate the N_2-fixation capacity of berseem in field trials (Williams et al., 1990). The instant uptake of available soil N by the roots diminishes also the danger of aquifer pollution caused by the leaching of nitrate below the root system, provided the application of excess water is prevented by the judicious control of depth and frequency of irrigation.

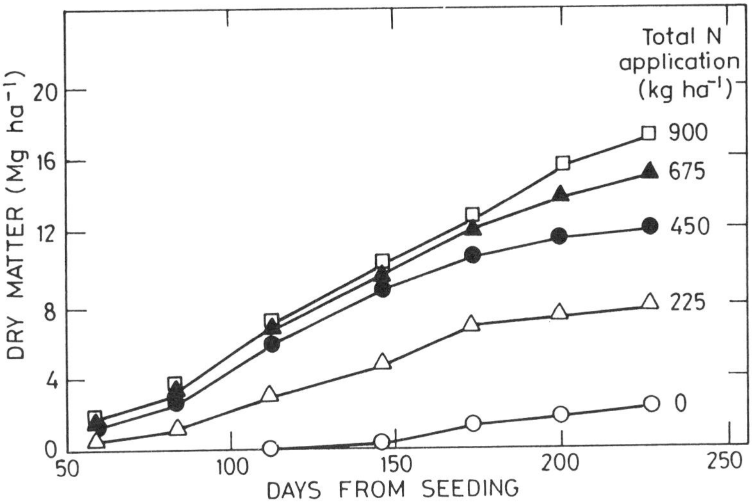

Fig. 9.3. Cumulative yields from harvests at monthly intervals, of irrigated annual ryegrass fertilized with various rates of N (as ammonium nitrate) in five equal applications (at seeding and after each of the first four cuttings). Seeding in October at Brawley, California. Summer-seeded Sudan grass preceded ryegrass (Hagemann et al., 1984).

Crude protein content of the forage rises markedly with the increase in fertilizer N. The application rates required for maximal DM yield cause a steep rise in the NO_3-N of the forage to levels that cause acute poisoning of livestock, when used as the sole or primary source of feed (Hagemann et al. 1984; Leshem, 1976). Forage with NO_3-N levels up to 4 g kg^{-1}, on a dry weight basis, is deemed safe when its contribution to the total DM intake of animals is no more than approximately 25%. Most experiments show that NO_3-N levels fluctuate during the growing season, with a peak in spring. An application of 120–140 kg N ha^{-1} at planting and after each harvest is expected to maximize forage yield without causing plant NO_3-N concentration to exceed animal tolerance. Toxic NO_3-N levels of grazed ryegrass are quoted by Lyons et al. (1992).

Phosphorus and Potassium

Annual ryegrass has no specific P and K requirements. Usually, fertilizer P is applied depending on the availability of soil P. Most of the soils in semi-arid and arid zones contain sufficient available K to sustain intensive growth of high-N-fertilized AR. In some cases, however, it has to be applied to maintain high forage production.

Hay

The relatively high moisture contents of the forage during vegetative growth, coinciding with periods of rain are serious obstacles in the harvesting and curing of hay. Delaying the date of seeding until late autumn may partly overcome these problems, but this delay will result in lower total forage yields compared with sowing at the end of the summer. During the autumn to early spring period, the evaporative demand of the atmosphere is relatively low, even under desert conditions, and causes a slow rate of water loss from the plants' surface. Harvesting should be carried out when the fresh mass does not exceed approximately 3 Mg ha^{-1}. The use of forage harvesters that condition the cut material by crimping the stems between rollers, spreading the cut herbage in thin layers, teddering and preventing densely packed windrows will improve the circulation of air through the leaf mass and thereby hasten the rate of curing. Rapid curing of hay before baling does not only ensure optimum forage quality but also diminishes risks posed by weather conditions.

Common haymaking procedures can be followed during the spring, when plant DM and the evaporation rate are considerably higher than during the winter.

Silage

Pre-wilting or the addition of chemicals are essential for successful fermentation, especially when high-moisture grass is ensiled. These measures are needed to keep seepage and other storage losses to a minimum.

The place of AR in forage programs in Algeria is shown in Figure 9.4.

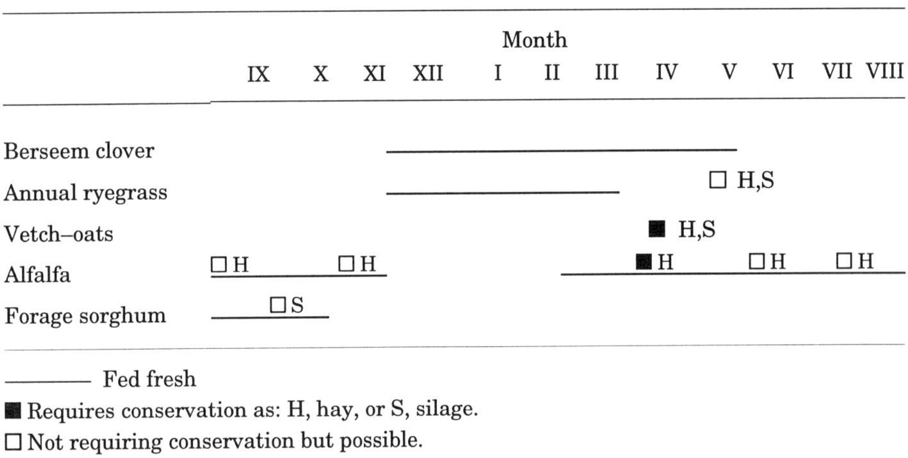

Fig. 9.4. Sequence of irrigated forage crops, including annual ryegrass, and their utilization in feeding programs in Algeria. (Adapted from Ahim et al., 1975).

9.9 CONCLUSIONS

This cool-season grass is capable of producing high forage yields with excellent nutritional value and offers ample utilization options. For optimum production, annual ryegrass does best on permeable soils of medium to high fertility. Even short periods of waterlogging affect stand persistence adversely.

Evapotranspiration and dry matter yield were found to be closely related. The shallow root system is a typical characteristic of this crop and makes it particularly likely to develop soil-water deficits rapidly. Under conditions of high evaporative demand, irrigation may be required once in 7–10 d depending on soil type. Neither dry matter yields nor forage quality decreased with saline water concentrations below approximately 6 dS m^{-1} when annual ryegrass was grown on a loessial soil in the northern Negev. Sewage water supplied sufficient nutrients for obtaining optimal forage yields.

Maximum dry matter and protein yields require copious amounts of mineral nutrients, specifically N, but excessive uptake of N can increase nitrate in the forage to levels poisonous to livestock.

REFERENCES

Ahmin, M., Kolli, R. and Lemaire, G. 1975. Rendement et valeur alimentaire de cinq varieties de Ray-grass d'Italie cultivées en Mitidja en relation avec le rythme d'exploitation. Fourrages 63: 35–45.

Aitken, Y. 1985. Temperate herbage grasses and legumes. pp. 185–202. In: A.H. Levi (Ed.) Handbook of Flowering, I. CRC Press, Boca Raton, FL.

Ameziane, T.E. 1979. Croissance et productivité du ray-grass d'Italie en zone méditerranéenne irrigée: Comparaison avec le trefle d'Alexandrie. Fourrages 78: 103–122.

Bartholomew, P.W. and Williams, R.J. 1981. Nitrogen application for short-term forages in Saudia Arabia. I. Winter annuals: wheat, oats and Italian ryegrass. Expl. Agric. 17: 271–279.

Douglas, J.T., and Crawford, C.E. 1989. Soil compaction and novel traffic systems in ryegrass grown for silage: Effects on yield, quality and nitrogen uptake. 16th Int. Grassland Conf. (Nice, France), pp. 175–176.

Dovrat, A. and Lachover, D. 1958. Yields and forage value of ryegrasses, prairie grass and oats under irrigation in comparison with berseem (Egyptian clover). Ktavim 8: 295–306.

Farnworth, J. and Said, S.A. 1983. The response of irrigated RVP Italian ryegrass to phosphate, potash and varieties. Dhamar Agr. Impr. Center 75: 3–5.

Garcia-Criado, L., Garcia-Ciudad, A. and Garcia-Criado, B. 1984. Irrigated temporary grasslands of mixed pasture species. An. Cent. Edafol. Biol. Apl. CSIC, Salamanca 10: 167–183.

Garwood, E.A. and Sinclair, J. 1979. Use of water by six grass species. 2. Root distribution and use of soil water. J. Agric. Sci. Camb. 93: 25–35.

Goodenough, D.C.W., Macdonald, C.I. and Morrison, A.R.J. 1984. Growth pattern of Italian ryegrass cultivars established in different seasons. J. Grassl. Soc. South Afr. 1: 21–24.

Goren, A. and Kessler, E. 1980. Ryegrass for hay. Israel Agric. Ext. Serv. Leafl. (in Hebrew). 2 pp.

Graves, W.L. and Munoz, F. 1987. Economic considerations for the conversion of annual ryegrass winter pastures to berseem clover in the Mexicali Valley. Coop. Ext. Univ. of Calif., San Diego, CA. 1 p.

Graves, W.L., Williams, W.A., Wegrzyn, V.A., Calderon, M.D., George, M.R. and Sullins, J.L. 1987. Berseem clover is getting a second chance. Calif. Agric. Sept.–Oct. pp. 8–16.

Hagemann, R.W., Ehlig, C.F. and Reynoso, R.Y. 1984. Soil nitrate level best measure of ryegrass nitrogen needs in Imperial Valley. Cal. Agric. Jan.–Feb. pp. 24–25.

Leshem, Y. 1976. The effect of applied nitrogen and cutting intervals on yield, nitrogen, and N-NO_3 content of ryegrass. Hassadeh 56: 1891–1897 (in Hebrew).

Leshem, Y., Amir, S., Halevi, A., Katz, Y., Tsuk, Y., Ashbel, G., Hefer, Y. and Kali, Y. 1989. Ryegrass: a quality forage. Hassadeh 69: 1575–1578 (in Hebrew).

Leshem, Y., De Malach, Y., Hofman, Z. and Yonathan, R. 1986. Cultivation of ryegrass for hay in the Negev region. Hassadeh 67: 1784–1790 (in Hebrew).

Leshem, Y., De Malach, Y. and Magnusson, D. 1989. Growing ryegrass with saline irrigation water. Hassadeh 69: 1784–1787 (in Hebrew).

Lowe, K.F., Bowdler, T.M., Ostrowski, T.M. and Stillman, S.L. 1983. Comparison of the yield, nitrogen and phosphorus content, and rust infection of irrigated ryegrass swards in south eastern Queensland. Aust. J. Exp. Agric. Anim. Husb. 23: 294–301.

Lowe, K.F. and Hamilton, B.A. 1985. Dairy pastures in the Australian tropics and subtropics. Proc. 3rd Aust. Conf. Trop. Pastures, Trop. Grassl. Soc. Austr. Occ. Publ. 3.

Lyons, D.J., Hawley, G.M. and Jeffrey, A.T. 1992. A sap test for the early diagnosis of toxic levels of nitrate in ryegrass. Trop. Glassl. 26: 165–170.

Maas, E.V. 1985. Crop tolerance to saline sprinkling. Plant Soil 89: 273–284.

Mejjati-Alami, M., Rihani, N. and Guessous, F. 1989. Composition chimique, valeur alimentaire et production du ray-grass d'Italie (*Lolium multiflorum*) conduit en irrigué sous climat méditerranéen. Proc. 16th Int. Grassl. Congr. (Nice, France) pp. 831–832.

Norris, I.B. 1982. Soil moisture and growth of contrasting varieties of *Lolium, Dactylis* and *Festuca* species. Grass Forage Sci. 37: 273–283.

Norris, I.B. 1985. Relationship between growth and measured weather factors among contrasting varieties of *Lolium*, *Dactylis* and *Festuca* species. Grass Forage Sci. 40: 151–159.

Penman, H.L. 1970. Woburn irrigation, 1960-1968. J. Agric. Sci. Camb. 75: 75–88.

Reyne, Y., Garambois, X., Angevain, J. and Gaubert, J.L. 1977. Note sur la valeur alimentaire en zone méditerrannéénne irriguée du ray-grass d'Italie et du sainfoin Fakir distribués en vert. Fourrages 69: 85–96.

Robbins, G.B. and Bushel, J.J. 1987. Productivity of grazed irrigated pastures of ryegrass, berseem clover and ryegrass–berseem clover in south-east Queensland. Trop. Grassl. 21: 133–138.

Robbins, G.B., Bushy, G.J. and Faulkner, G.B. 1980. Annual production from annual pastures of irrigated ryegrass in the Burnett district of south-east Queensland. Trop. Grassl. 14: 40–43.

Robson, M.J., Ryle, G.J.A. and Woledge, J. 1988. The grass plant — its form and function. pp. 25–83. In: M.B. Jones and A. Lazenby (Eds.) The Grass Crop. Chapman and Hall, London, UK.

Rosello, B. 1971. Primeros ensayos realizados con el Ray-grass Westerwold en el valle del Ebro. An. INIA Ser. Prod. Anim. 1.

Rotili, P., Gnocchi, G., Scotti, C. and Serena, A. 1985. Bio-agronomic evaluation of *Lolium multiflorum* populations collected in irrigated plains of northern Italy. Proc. 15th Int. Grassland Congr. (Kyoto, Japan), pp. 140–141.

Sagi, M. 1992. M.Sc. thesis (unpublished), Hebrew University of Jerusalem (in Hebrew).

Simon, J.C. 1981. Contribution à l'étude ecophysiologique de la fase semis-levée du raygras d'Italie (*Lolium multiflorum*). I. Etude en conditions controlées de l'influence du facteur thermique. Agronomie 1: 339–344.

Simon, J.C. and Lemaire, G. 1987. Tillering and leaf area index in grasses in the vegetative phase. Grass Forage Sci. 42: 373–380.

Skjelvag, A.O. 1986. Temperature and physiological development of Westerwolth ryegrass. Forsk. Fors. Landbruhet 37: 219–224 (in Norwegian).

Smith, H.R.H., Dempsy, C.P. and Schaffner, M. 1988. Identification of companion small grains for Midmar Italian ryegrass. J. Grassl. Soc. South Afr. 5: 150–153.

Thomas, H. 1986a. Water use characteristics of *Dactylis glomerata* L., *Lolium perenne* L. and *L. multiflorum* Lam. plants. Ann. Bot. 57: 211–223.

Thomas, H. 1986b. Effect of rate of dehydration and leaf water status and osmotic adjustment in *Dactylis glomerata* L., *Lolium perenne* L. and *L. multiflorum* Lam. Ann. Bot. 57: 225–235.

Tisserand, J.L. and Alibes, X. 1989. Feeds of the Mediterranean area. pp. 305–323. in: R. Jarrige (Ed.) Ruminant Nutrition: Recommended Allowances and Feed Tables. INRA, Paris.

Whyte, R.O., Moir, T.R.G. and Cooper, J.P.C. 1959. Grasses in Agriculture. FAO Agric. Stud. 42. 404 pp.

Williams, W.A., Graves, W.L. and Cassman, K.G. 1990. Nitrogen fixation by irrigated berseem clover versus soil nitrogen supply. J. Agric. Crop Sci. 3: 202–207.

Chapter 10

FODDER BEET

10.1 INTRODUCTION

Fodder beet, stock beet and mangel are common names given to the extremely versatile species, *Beta vulgaris*, which also includes sugar beet and table beet. All these crops, each with a wide range of cultivars, have the common characteristic that their leaves and fleshy roots are used, rather than their reproductive parts. In the temperate countries of Europe, fodder beet is usually seeded in the spring and harvested in late autumn. It is either fed fresh or as a silage, or used as silage component during the cold winter months. The high yielding ability, favorable lactogenic properties and outstanding water-use efficiency of fodder beet combine to make it a potentially important forage crop in dry regions (Arnon, 1972). In the mild-winter areas of the Mediterranean basin, fodder beet is grown as a winter annual and harvested from the early spring into the summer. In Israel, the feeding of fresh beets to dairy cattle, especially on smallholder farms, was common practice until a few years ago. The reasons for fodder beet's limited distribution are the lack of sufficiently developed technologies in weed control, mechanization, and preservation. Renewed interest in this crop has been expressed recently in Egypt (Abou El-Hassan et al., 1971; Rammah et al., 1984; Shalaby et al., 1989a,b; Nowar et al., 1989), Cyprus (Hadjichristodoulou, 1976), Yemen (Farnworth and Said, 1982), Tunis (Kopp, 1976) and Israel (Leshem, 1992).

Recent reviews of the production and feeding of fodder beet in northern Europe are provided by Larsen (1985) and Pollott (1990). Information on beet diseases and insects has been compiled by Whitney and Duffus (1986) and by Benada et al. (1987).

Members of cultivated *Beta* are closely related to the wild sea-beet *B. vulgaris* spp. *maritima*, which today is found in coastal habitats along the Mediterranean Sea and the Atlantic coast, as well as on inland steppes.

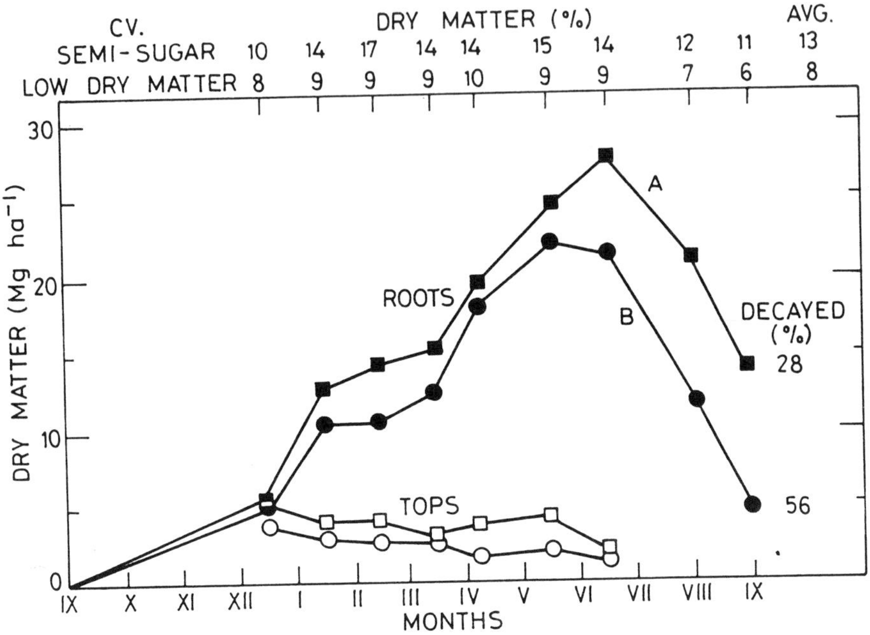

Fig. 10.2. Typical curves of dry matter accumulation for roots and tops (leaf blades + petioles) in autumn-seeded fodder beet: (A) semi-sugar, and (B) low-DM cultivars. Number of plants 60×10^3 ha^{-1}. Seeding first week of September. Means of 4 years. Bet Dagan, Israel. (From Dovrat and Leshem, 1965).

zero and even became negative, because leaf loss through senescence exceeded the growth of new leaves. Dry matter yield and DM content of roots diminished dramatically with the onset of higher temperatures from June onwards, most probably as the result of (i) increased maintenance respiration, (ii) cessation of leaf growth, and (iii) root decay.

Significant yield differences were found between cultivars differing in DM content in variety trials in Israel (Fig. 10.2). The high-DM cultivars (i.e. semi-sugar cultivars) with an average 5-percentage units higher DM content, yielded almost 15% more total DM per hectare than low-DM cultivars. However, the total fresh weight yields of the low-DM cultivars were higher than those of the high-DM cultivars. Additionally, the decrease in yield due to decay of roots during the hot summer months was almost twice as great in the low than in the high-DM cultivars. These results indicate clearly the advantage in using high-DM cultivars when the harvesting of roots is scheduled to extend into the months with high temperatures. It is usual practice in Israel to sow low-DM cultivars at higher plant density, for harvesting in the spring, and high-DM cultivars for harvesting in late spring and early summer.

Prolonged periods of low temperature followed by higher temperatures and longer day length may cause elongation of stems (bolting). The bolting of plants is associated with reduced root thickening and low forage quality. Cultivars differ in bolting resistance.

10.3 ESTABLISHMENT

Cultivation aspects of the sugar beet crop, which are similar to those of fodder beet, were reviewed by Johnson et al. (1971) and Arnon (1972). Some of the aspects pertinent to fodder beet are mentioned below.

10.3.1 Soil Requirements

Fodder beets prefer deep, light-textured, fertile and well-drained soils with pH 6–7. In the rotation sequence they are sown once in 4–5 years, in order to evade problems caused by nematodes and other soilborne pests.

10.3.2 Seedbed Preparation

The final stages are designed to eliminate germinating weed seeds and to create a soil surface environment conducive to precision planting, rapid germination and emergence of seedlings. Bedding will improve seedling development and, if required, the maintenance of more definite furrows.

Weed Control

The fodder beet crop is severely affected by weed competition, because it takes a long time to cover the ground and become competitive. Sugar beet growers have recently changed over to the use of repeated low-dose post-emergence herbicide programs. This practice results in better weed control, reduced chemical use and reduced crop damage (Cromack, 1985). Many of the weed problems can, of course, be circumvented by *in situ* planting of seedlings.

10.3.3 Date and Methods of Seeding

Fodder beet can be seeded throughout the autumn and is usually staggered in order to extend the harvesting season. When determining the seeding date in cotton growing areas, growers should take into account the incidence of attacks by *Prodenia*. The preparation of seedlings in a nursery and subsequent planting in the field, vs. *in situ* seeding, has the following important advantages: (i) saving of irrigation water, (ii) selection of sturdy seedlings, (iii) better control of planting density, and (iv) easier weed control. Planted seedlings may develop more lateral roots than seedlings growing *in situ*, and may render root lifting more difficult.

Precision planting is facilitated by the availability of seeds which are uniform in shape and size (achieved by pelleting, decortication, or the use of monogerm seed) and of suitable drills. Uniform spacing within rows is desirable, particularly when machine harvested, although roots have the ability to compensate for wide variations in density, provided that there are no large gaps. Total DM yield per ha need not necessarily be much affected by the lack of uniformity, but the size of the individual roots may vary considerably. Care should be taken, however, to prevent the ingestion of small beets by animals, because of the risk to them of blockages in the throat, which may lead to suffocation. The recommended density is 70,000–100,000 plants per ha, depending on planting date, cultivar and field conditions. Between-row spacing may be adjusted to the current sowing distance of other crops, irrigation and harvest operations on the same farm.

10.3.4 Cultivars

A great number of cultivars are available differing in shape, color, level of ploidy, root branching and DM content (OECD, 1991). For practical purposes they are grouped according to: (i) the DM content of the root; and (ii) whether the 'seeds' contain one viable seed only (monogerm) or are fruits (clusters) containing more than one seed (multigerm) (Descriptive List of Varieties, 1991). Seed-certifying agencies in Europe distinguish between cultivars having roots (when harvested in the autumn) with (i) low DM content (100–140 g kg^{-1}), and usually globular-shaped; and (ii) high DM (140–180 g kg^{-1}) and usually oval-shaped and tapered. The DM content of these cultivars depends greatly on the time of harvesting and the soil water regime, with the result that under irrigated conditions the DM content of the roots is quite variable. This explains why the DM contents of similar cultivars in different experiments can differ quite considerably (Dovrat and Leshem, 1965; Hadjichristodoulon, 1976; Samsky and Dar, 1986; Nowar et al., 1989).

10.4 MINERAL NUTRITION AND FERTILIZER USE

Most of the fertilizer recommendations for irrigated fodder beets are based on those for sugar beets, because of the similarity in demand for nutrients for development and growth. Both fodder and sugar beets require a constant supply of relatively high amounts of readily available soil N, P and K in order to support maximum growth rates of tops and roots during most of the growing season. However, in view of the high potential DM yields of fodder beets, it may be expected that their demand for nutrients exceeds that of sugar beets. Mineral nutrition and the yield response of sugar beets to fertilizers were studied extensively and reviewed by Hills and Ulrich (1971) and by Schmehl and James (1971). Critical concentrations of N, P and K in petioles of leaves are shown in Table 10.1.

TABLE 10.1

Critical concentrations of nutrients in petioles from recently matured leaves of sugar beet plants. At these concentrations growth was retarded by 10%. (Adapted from Ulrich and Hills, 1969)

Nutrient	Critical concentration
Nitrogen	304 mg kg^{-1}
Phosphorus	240 mg kg^{-1}
Potassium (when Na >15 g kg^{-1})	10 g kg^{-1}

Nitrogen

Large amounts of N, often applied in split applications, are essential for adequate top and root growth and are estimated at 200–325 kg N ha^{-1}. Excessive rates of N can raise NO_3 concentration in the roots to levels toxic to animals.

Potassium

Experiments with fodder beet, carried out in New Zealand (Magat and Goh, 1988) and in Britain (Cromack, 1985), showed that fertilization with NaCl and KCl increased significantly the yields of both fresh and dry roots, and that total K uptake was the main nutritional factor determining these yield increases.

10.5 CROP WATER REQUIREMENTS AND IRRIGATION

It is assumed that the water requirements of fodder beet are similar to those of sugar beet, although fodder beet varieties are selected for low sucrose concentration and high fresh (dry) weight root yield, and sugar beet varieties for high sucrose concentration and lower fresh weight root yields. Irrigation practices may differ, however, since fodder beets are sown almost exclusively in autumn and, in contrast to sugar beets, are harvested intermittently from early spring to mid summer. Irrigation practices of both crops have in common that they are directed toward rapid stand establishment, early attainment of a full leaf canopy and avoidance of periodic water stress. Plant–water relationships and irrigation of sugar beets are reviewed by Loomis and Haddock (1967), Jensen and Erie (1971), Martin (1983) and Hills et al. (1990).

10.5.1 Water-Production Functions

Evapotranspiration

Germination and seedling development of fodder beet proceed slowly, taking 2–3 months until a full canopy of leaves is attained. Because of the relatively low plant population density, the bare soil surface is exposed over a relatively long period to water loss by evaporation before a full cover is reached. Useful means to cut down on water loss by evaporation during this period are to seed at a higher plant density and, more specifically, to plant seedlings instead of seeding *in situ*.

No reports are available on ET measurements of fodder beet. Nevertheless, the few available measurements of autumn-seeded sugar beets may assist in estimating the water use of fodder beet, assuming that their evaporative demands are similar. Average seasonal ET of non-stressed sugar beets was approximately 1050 and 1200 mm in Arizona and southern California, respectively (Erie and French, 1968; Ehlig and LeMert, 1979). Rainfall is sporadic in both regions. During the period of rapid root growth in the Phoenix area (Arizona), the average K_c was approximately 0.9 but lower in southern California, possibly as a result of different methods of measurements.

10.5.2 Sensitivity to Water Stress

Moisture sensitivity is especially noticeable during the early establishment stage, i.e., germination, emergence, and seedling growth (until first true leaves). The major causes of this sensitivity are the relatively small seed size and the slow growth rate of the seedlings. No other periods of sensitivity to moisture stress have been reported, since the fodder beet crop usually remains vegetative throughout its growth cycle. Of course, periodic soil water deficits during the growing season will retard root DM accumulation, because of restricted leaf expansion and delays in the initiation of new leaves. However, as long as the roots have not reached maturity (i.e., in mid summer), replenishing soil water deficits is expected to restore normal root weight accumulation.

10.5.3 Criteria for Irrigation Scheduling

There is general agreement that 60–70% of the available water in the root zone of sugar beets can be depleted without yield (or sucrose) loss, if each irrigation supplies the amount depleted (Hills et al., 1990). This may apply also to fodder beets unless and until results from experiments show otherwise. It also requires verification of whether roots of fodder beets growing in well-drained soils can reach 1.8 m deep or more, and whether most of the water uptake is from the top 0.9 m, as in sugar beets.

10.5.4 Sensitivity to Low Water Quality

Salinity

A well-established sugar beet plant may easily tolerate soils with electrical conductivity values near 8–10 dS m^{-1}, but germination may be appreciably delayed and reduced at 6 dS m^{-1} (Loomis and Haddock, 1967; Maas, 1984). Methods used under saline conditions which avoid salt accumulation near the young seedlings are described by Reeve and Fireman (1967).

10.6 HARVEST MANAGEMENT

During the harvesting period from early spring to summer, the roots increase in weight while the amount of foliage decreases. When roots reach maturity, the leaves have usually dried and been shed. While it is not possible to store roots for any useful length of time under warm and dry climatic conditions, it is desirable to lift a few days' ration at a time. Fodder beets, if required, can be processed in feed centers, where complete feed rations are daily prepared to feed the herd or are distributed to other farms. After field harvesting and hauling to the center, the roots are mechanically shaken to free them from adhering (dry) soil and, after chopping, mixed with the rest of the feed ration. Chopped roots, in addition to their high feed value, are also a desirable wetting agent in the ration (Samsky and Dar, 1986; Tsur et al., 1986; Shefet, 1990). The roots can also be strip grazed *in situ* without waste, using an electric fence to delimit a day's ration to the grazing animals.

Harvesting system and cultivar selection are closely linked. In most low-DM cultivars, at least half of the root is above the ground. This characteristic means that the cultivars are best suited for hand lifting or top-lifting types of harvester. They are also well suited to heavier soils, where soil adhesion to the below-ground part is more of a problem. Most hybrid high-DM cultivars have a uniform root shape, with approximately two-thirds of the root in the ground, which makes them suitable for lifting with share-type harvesters. Beet harvesters, including those traditionally used (with adjustments) for the lifting of sugar beets, are described by Robinson (1985). New harvesting technology for the preparation of whole-crop fodder beet silage is reported by Meagher (1990).

Commercial root yields in Israel range between 23 and 42 Mg DM ha^{-1} (Tsur et al., 1986).

10.7 FORAGE QUALITY AND UTILIZATION

Fodder beet is considered an excellent source of concentrate because of its high IVDMD (0.75–0.85), high ME (11.8–13.4 MJ kg DM^{-1}), and low protein (60–75 g kg^{-1}) contents in the DM. The main energy sources are water-soluble carbohydrates, principally sucrose, glucose and fructose (Clark et al., 1987). In some north European countries fodder beet is sometimes used to replace concentrates in feed rations for high-yielding dairy cows in early lactation. Total DM intake is often increased when fodder beet is included in the ration. There is conflicting evidence on the effect of feeding fodder beet on milk yield, but most workers have shown that fodder beet improves milk quality, in terms of butterfat and milk protein content (Larsen, 1985; Fitzgerald, 1990; Roberts and Martindale, 1990).

The removal of adhering soil on roots, occurring especially when fodder beet is grown on heavy clay soil, requires proper attention. The level of soil contamination can influence the acceptability of the roots as feed. In addition, the feed value of the root depends on the amount of soil contamination, which is reflected in the ash content of the DM, and should not exceed approximately 50 g kg^{-1}. High ash contents can cause acute restraint of the microbial activity in the rumen (Larsen et al., 1964).

Variability in fodder beet DM concentration, and thus in nutritional value, due to growth stage, growing conditions and cultivar, is an inherent feature of this crop and should be taken into account when formulating feed rations.

Whole Crop Silage

The high yield potential of fodder beet can be exploited also by ensiling (chopped) roots plus tops as soon as the field reaches its maximum yield. Recent long-term experiments in Germany and Denmark have shown that the palatability and high feeding value of fodder beet can fully be preserved during the ensiling process. Practical experience established that mixing roots plus tops with approximately 12 kg of an absorbent (such as 12% NH_3– or 15% NaOH-treated straw) added to 100 kg of fresh material, will reduce the risk of effluent loss to a minimum (Kleinhout, 1990).

10.8 CONCLUSIONS

The high yielding capacity, lactogenic properties, high water use efficiency, and relatively long growing season of this high-energy root crop combine to make it a potentially useful irrigated forage crop. The further extension (in area) of this crop depends mainly on the development of weed control technologies and the mechanization of harvesting. Fodder beet is sensitive to saline conditions during germination, but is noted for its tolerance once the crop has become established.

Rapid stand establishment and early attainment of a full leaf canopy in the autumn, reduce water loss by soil evaporation. Alternatively, the planting of seedlings saves irrigation water and at the same time helps to control weeds. Soil water deficits beyond evapotranspirational demands decrease the rate of dry matter accumulation.

High dry matter cultivars with roots protruding approximately one-third above the ground facilitate machine harvesting. Soil contamination reduces the feed value of the roots. Harvesting, usually at few days' intervals, can continue until roots reach full maturity during the summer, after which they lose weight because of increased respiration and root decay. Excessive N fertilization can increase the nitrate concentration in roots to levels poisonous to livestock.

REFERENCES

Abou El-Hassan, A., El-Samman, S., Radwan, A. and Stino, G. 1971. The nutritive value of fodder beet under the UAR conditions. J. Anim. Prod. 11: 233–243.

Arnon, I. 1972. Crop Production in Dry Regions. II. Systematic Treatment of the Principal Crops. pp. 599–601. Leonard Hill, London.

Artschwager, E. 1926. Anatomy of the vegetative organs of the sugar beet. J. Agric. Res. 33: 143–176.

Benada, J., Sedivy, J. and Spacek, J. 1987. Atlas of Diseases and Pests in Beet. Elsevier Science Publishers, Amsterdam, The Netherlands. 272 pp.

Clark, P., Givens, D.I. and Brunnen, J.M. 1987. The chemical composition, digestibility and energy value of fodder-beet roots. Anim. Feed Sci. Technol. 18: 225–231.

Cromack, H.T.H. 1985. Fodder beet growing the crop. Fodder Beet — Its True Potential. National Agriculture Conf., (Kenilworth, UK). 1–8 pp.

Descriptive List of Varieties of Field Crops. 1991. Centrum Rassenond. Zaadtechn. Wageningen, The Netherlands. pp. 171–172 (in Dutch).

Dovrat, A. and Leshem, Y. 1965. Semi-sugar beets. Hassadeh 45 (5–8): 1253–1256

Erie, L.J. and French, D.F. 1968. Water management of fall-planted sugar beets in Salt River Valley of Arizona. Trans. Am. Soc. Agric. Engrs. 11: 792–795.

Esau, K. 1977. Anatomy of Seed Plants. John Wiley & Sons, Chichester, UK. 550 pp.

Ehlig, C.F. and LeMert, R.D. 1979. Water use and yields of sugarbeet over a range from excessive to limited irrigation. Soil Sci. Soc. Am. J. 43: 403–407.

Farnworth, J. and Said, S.A. 1982. A comparison of fodder brassica and root crops in 1981 and 1982 grown under irrigation. Publ. Dhamar Agric. Impr. Centre. Yemen No. 44: 7 pp.

Fick, G.W., Loomis, R.S. and Williams, W.A. 1978. Sugar beet. pp. 259–295. In: L.T. Evans (Ed.) Crop Physiology. Cambridge University Press, Cambridge, UK.

Fitzgerald, S. 1990. Feeding fodder beet to replace a concentrate supplement or as an additional feed in the diet of lactating cows. pp. 173–178. In: G.E. Pollott (Ed.) Milk and Meat From Forage Crops, Brit. Grassland Soc. Occ. Symp. 24: 173–178.

Hadjichristodoulou, A. 1976. Trials of dryland fodder beet in Cyprus. Exp. Agric. 12: 61–67.

Hills, F.J., and Ullrich, A. 1971. Nitrogen nutrition. pp. 111–136. In: R.T. Johnson, J.T. Alexander, G.E. Rush and G.R. Hawkes (Eds.) Advances in Sugarbeet Production. Iowa State University Press, Ames, IA.

Hills, F.J., Winter, S.R. and Henderson, D.W. 1990. Sugarbeet. In: D.E. Stewart and D.R. Nielsen (Eds.) Irrigation of Agricultural Crops. ASA, CSA and SSSA Monogr. 20: 795–810.

Jensen, M.E. and Erie, L.J. 1971. Irrigation and water management. pp. 189–222. In: R.T. Johnson, J.T. Alexander, G.E. Rush and G.R. Hawkes (Eds.) Advances in Sugarbeet Production. Iowa State Univ. Press, Ames, IA.

Johnson, R.T., Alexander, J.T., Rush, G.E. and Hawkes, G.R. (Eds.) 1971. Advances in Sugarbeet Production: Principles and Practices. Iowa State Univ. Press, Ames, IA. 470 pp.

Kleinhout, A. 1990. The use of fodder beet as a whole crop: Experience in Denmark and elsewhere. pp. 157–172. In: G.E. Pollott (Ed.) Milk and Meat from Forage Crops. British Grassland Soc. Occ. Symp. 24.

Kopp, E. 1976. Production potential of the semi arid region of Haute Vallée de la Mesjera in Tunisia under sprinkler irrigation. Ann. Inst. Nat. Rech. Agron. Tunis. 49: 356. (in French).

Larsen, J.B. 1985. The practice and new development of the production and feeding of fodder beet in Denmark. Fodder Beet — Its True Potential. Nat. Agr. Conf., Kenilworth, UK. pp. 15.

Leshem, Y. 1992. Fodder beet: a crop of renewed interest. Hassadeh 72: 836–840 (in Hebrew).

Loomis, R.S. and Haddock, J.L. 1967. Sugar, oil and fiber crops. Part 1: Sugar beets.

In: R.M. Hagan, H.R. Haise, T.W. Edminster (Eds.) Irrigation of Agriculture Lands., ASA Monogr. 11: 640–648.

Loomis, R.S. and Nevins, D.J. 1963. Interrupted nitrogen nutrition effects on growth, sucrose accumulation and foliar development of the sugar beet plant. J. Am. Soc. Sugarbeet Technol. 12: 309–322.

Maas, E.V. 1984. Salt tolerance of plants. pp. 57–75. In: B.R. Christie (Ed.) Handbook of Plant Science in Agriculture, Vol. 2. CRC Press Inc., Boca Raton, FL.

Magat, S.S. and Goh, K.M. 1988. Effects of chloride fertilizers on yield and uptake of chloride, potassium and sodium by fodder beet (*Beta vulgaris* L.) in two New Zealand soils. J. Agric. Sci. 111: 207–216.

Martin, S.S. 1983. Sugarbeet. pp. 423–443. In: I.D. Teare and M.M. Peet (Eds.) Crop–Water Relations. John Wiley & Sons, New York, NY.

Meagher, J. 1990. Fodder beet: A fodder source of concentrate. pp. 179–184. In: G.E. Pollott (Ed.) Milk and Meat from Forage Crops. Brit. Grassland Soc. Occ. Symp. 24.

Morton, A.B. and Watson, D.J. 1948. A physiological study of leaf growth. Ann. Bot. N.S. 12: 281–310.

Nowar, M.S., Abd El-Baki, S.M., Marie, I.F. and Zaki, A.A. 1989. Evaluation of new varieties of fodder beet (Majoral and Polyproductiva) as new summer forages in Egypt. pp. 859–860. Proc. 16th Int. Grassland Congr. (Nice, France).

OECD. 1991. List of Cultivars Eligible for Certification. OECD, Paris. 90 pp.

Pollott, G.E. (Ed.) 1990. Milk and Meat from Forage Crops. pp. 137–190. British Grassland Soc. Occ. Symp. 24.

Rammah, A.M., Ali, F.M. and Hassan, M.T. 1984. Evaluation of fodder beet cultivars for different locations and years. EMCIP Pub. 1: 215–224.

Reeve, R.C. and Fireman, M. 1967. Salt problems in relation to irrigation. In: R.M. Hagan, H. R. Haise and T.W. Edminister (Eds.) Irrigation of Agricultural Lands. ASA Monogr. 11: 988–1008.

Roberts, D.I. and Martindale J.F. 1990. Fodder beet: A review of research findings in relation to animal production. pp. 137–152. In: G.E. Pollott (Ed.) Milk and Meat from Forage Crops. British Grassland Soc. Occ. Symp. 24.

Robinson, P.J. 1985. Fodder beet harvesting and storage. In: Fodder Beet — Its True Potential. Proc. Nat. Agric. Conf. (Kenilworth, UK). 15 p.

Samsky, J. and Dar, J. 1986. Fodder beet for the feed center. Hassadeh 67: 235–237 (in Hebrew).

Schmehl, W.R. and James, D.W. 1971. Phosphorus and potassium nutrition. pp. 137–169. In: R.T. Johnson, J.T. Alexander, G.E. Rush and G.R. Hawes (Eds.) Advances in Sugarbeet Production. Iowa State University Press. Ames IA.

Shalaby, A.S., Rammah, A.M., Abdul-Aziz, G.M. and Beshay, M.G. 1989a. Fodder beet: a new forage in Egypt. 1. Productivity and chemical analysis of some fodder beet (*Beta vulgaris* L.) cultivars sown at different locations in Egypt. Proc. 3rd Egypt–British Conf. on Anim. Feed and Poultry Prod. (Alexandria, Egypt), Vol. 1, pp. 133–143.

Shalaby, A.S., Abdul-Aziz, G.M., Rammah, A.M. and Beshay, M.G. 1989b. Fodder beet: a new forage in Egypt. 2. Nutritional evaluation of fresh fodder beet (*Beta vulgaris* L.) and its silages, with or without roughages. Proc. 3rd Egypt–British Conf. on Anim. Feed and Poultry Prod. (Alexandria, Egypt), Vol. 1, pp. 145–152.

Shefet, G. 1990. The feeding system for ruminants in Israel. Hassadeh Quart. 1: 78–80.

Terry, N. 1968. Developmental physiology of suger beet. I. Influence of light and temperature on growth. J. Exp. Bot. 19: 795–811.

Tsur, Y., Goldman, A. and Dar, Y. 1986. Cultivation of fodder beets and their use in the feed center. Gan Sade Umeshek: 25–27 (in Hebrew).

Ulrich, A. and Hills, F.J. 1969. Sugarbeet Nutrient Deficiency Symptoms — A Colored Atlas and Chemical Guide. Univ. Calif. Div. Agric. Sci.

Watson, D.J. and Baptiste, E.C.D. 1938. A comparative physiological study of sugar beet and mangold with respect to growth and sugar accumulation. I. Growth analysis of the crop in the field. Ann. Bot. 2: 437–480.

Whitney, E.D. and Duffus, J.E. 1986. Compendium of beet diseases and insects. Ann. Phytopath. Soc. Press. St Paul, MI. pp. 76.

Chapter 11

ECONOMIC EVALUATION OF FORAGE PRODUCTION OPTIONS ON MIXED FARMS

A. Goldman

11.1 INTRODUCTION

In certain cases the economic evaluation of forage crops is problematic. The profit of an economic activity is calculated by the difference between costs and revenues. Calculation of forage production costs does not differ from any other calculation. The calculation of income from a marketable forage activity (like hay, which has a market and a price) is relatively simple. However, the economic evaluation and assessment of income and profit from a non-marketable forage activity in a mixed farm (such as pasture, green-chop, silage or even hay which has no market or market price) is more complicated, for several reasons:

(i) the contribution is indirect (through the output of farm livestock in a different branch, eating the forage) and therefore a knowledge is needed of the precise feeding value, which comprises many components — energy and protein are the most significant economically;

(ii) as an auxiliary activity, the demand for forages is derived (quantitatively and qualitatively) from the alimentary needs of the animals throughout the year, while both excess and shortage are undesirable;

(iii) alternative forage crops may be used in any season, each having its specific cost and feed components content; purchased concentrates, by-products, etc., may replace the home-grown forage; and

(iv) it is possible to utilize forage crops in various ways — as pasture, or after cutting, fed fresh or conserving (hay, silage, etc.). The form in which forage is utilized greatly influences the cost and the feeding value of each forage activity.

Selection of the optimal combination of forage crops, their uses and the purchased feeds will increase the profitability of the farm. On a farm where cash-crops are also grown, optimization of the forage crops combination is further complicated, for each forage crop must compete in two cycles: (i) with other forage crops and with purchased feeds to be incorporated into the animals' ration, and (ii) with cash crops for the utilization of land and water (Fig. 11.1).

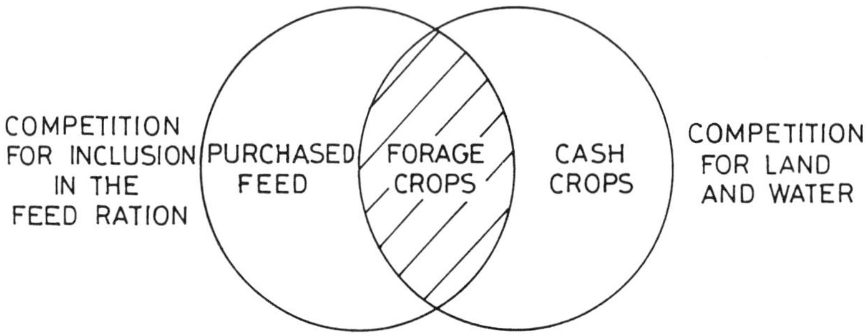

Fig. 11.1. Competitive factors in the growing of forage crops on the mixed farm.

The farmer needs decision-making tools for selecting crops and technologies for optimum stock feeding. Linear programming is today an accepted tool for handling such problems. This computerized mathematical method is widely used for the selection of feedstuffs. The user defines the limitations of the feed requirements for each group of animals, which set the minimum, equal or maximum level of each feed component (energy, protein, fiber, minerals, vitamins, etc.). For each source of feed supply, the user takes into account the content of the feed components required and the cost per ton. Using linear programming the optimal mix of feedstuffs to satisfy stock feeding requirements at minimum cost is determined. Linear programming is also used for the planning of field crops (at given restrictions of land, water, labor, machinery, etc.) and enables estimation of the maximum return from production inputs.

Although linear programming of the combined stock feeding–crops package would preclude the need for a separate economic evaluation of home-grown forages, it necessitates a very large matrix size. In the 1960s some Israeli kibbutzim (collective settlements) undertook such calculations, but they were discontinued because of complications in parameter definition and data provision to the computer.

11.2 COMPARATIVE CALCULATIONS

At the end of the 1970s a computer program to evaluate and compare different technological feeding options was developed in Israel by a team of extension and research workers familiar with forage-crop growing and livestock feeding (Goldman et al., 1981). The program was devised to calculate forage production and feeding costs. It enables a comparison between different technological alternatives (home-grown fresh or conserved forage, purchased feeds, industrial by-products, etc.) on a common basis, namely, the sum of all costs until the manger stage and the feeding value in the manger, according to the input costs and output values on a certain date. Costs related to land and capital outlay for irrigation were not included, since the system aims at guiding each grower in selecting between alternatives in his farm and not for comparison between farms.

The monetary value of home-grown forages not having a market price is expressed, as suggested by Morrison (1957), in terms of the monetary value of two major feed components (viz. energy and protein) in two widely used alternative feeds: soybean cakes (as the main supplier of protein) and cereal grain (as the main energy supplier). If one makes the simplifying assumption that the current market price of these two feeds represents the ratio between the supply and demand of protein and energy, it follows that:

(i) the price of one ton of corn = (value of one unit of protein) × (protein content of corn) + (value of one unit of energy) × (energy content of corn); and

(ii) the price of one ton of soybean cake = (value of one unit of protein) × (protein content of soybean cake) + (value of one unit of energy) × (energy content of soybean cake).

These two equations enable us to calculate the monetary value of one unit each of energy and of protein at a certain date. Under Israeli conditions, where the major part of the feed ration for high-yielding dairy cows consists of concentrates, and to which a minimum amount of required crude fiber as roughage is added, crude fiber can attain a monetary value. The price of vetch hay, a common marketable product, can be used for the calculation of the monetary value of a unit of crude fiber (when positive) as follows:

$$\text{C hay} - \frac{(\text{VCP hay} + \text{VEN hay})}{\text{CF hay}}$$

where VCF = value of 1 unit crude fiber; C hay = cost of 1 t hay; VCP hay = value of crude protein in hay; VEN hay = value of energy in hay; and CF hay = crude fiber content of hay.

The computer program provides the user with a number of outputs; examples of which are presented in Tables 11.1, 11.2 and 11.3.

– A list of all the relevant inputs and their normative prices at a certain date.

– Detailed calculation of the costs of each feed technology, which comprises all input elements summing up the total expenses per hectare (of forage crops) or per ton (of purchased feed), and their monetary value in the manger, according to energy, protein, crude fiber content, and their price per unit. An abbreviated calculation for corn silage, as an example, is presented in Table 11.1.

– A comparison (on a per-ton (Mg) basis) of different technologies from the point of view of the animal grower, is presented in Table 11.2.

– A similar comparison (on a per-hectare (ha) basis) from the point of view of the forage-grower is presented in Table 11.3.

In Tables 11.2 and 11.3 three levels of yields are considered, viz. high, intermediate and low, for two farm sizes: family farm (small, 10–50 cows) and collective farm (large, over 300 cows).

Table 11.1 lists the details of the cost of silage-corn activity production methods developed by extension workers directing the user to consider them when planning the crop. The overall water input in this calculation includes rain, although only irrigation water is "charged", according to its costs in different seasons. It is also possible to "charge" the use of rain water according to an alternative price (taken from the net income of a standard crop such as grain wheat), thus rendering more accurate the comparison with other technologies which also exploit rain water. Details of waste losses and treatments following the harvest enable greater precision in the comparison between different harvest technologies. In these calculations the value of crude fiber is included. The proportion of roughage in the total ration supplied to high-yielding dairy cows in Israel is relatively small, and generally dryland vetch hay is the main source of marketable crude fiber. If the price of hay exceeds the value of protein and energy it provides, then its crude fiber has an economic value. Usually fiber has no economic value, since the price of hay is lower than the values of its protein and energy.

Comparisons of production technologies of different crops and purchased feeds per Mg are given in Table 11.2. Three yield levels — high, intermediate and low (excluding crop failures) — are calculated. The correction coefficient for expenses for the lower levels (column 2) until harvest, is based on the assumption that not all the necessary inputs have been provided, and this is partially responsible for the low yields. Some extension workers claim that the use of a correction coefficient is not justified

TABLE 11.1

Example of the economic evaluation of a forage feeding activity (abbreviated calculation). Spring corn for silage, on fallow, per hectare (maximum fresh-matter yield, 100 t/ha with 28% DM, July 1980[t])

Input Description	Quantity	Unit	Price per unit ($)	Cost ($)
Irrigation water, Oct.–May	2500	m^3	0.11	277
Efficient rain	2700	m^3		
	Total costs to harvest			1132
Harvest losses	5.0	%		
Storage losses	11.5	%		
Silage making-equipment				66
Storage (facilities, materials, interest)				39
Silage feeding-equipment				36
Manpower for harvest and feeding	63.1	hour	3.63	23
	Total costs from harvest to manger			1648
	Total costs to manger			2780

Output Description	Quantity	Unit	Price per unit ($)	Total units	Value ($)
Forage in manger[z]	84.1	Mg			
Dry matter in forage	26	%			
Metabolic energy value[v]	2400	MgCal[x]	0.689	52500	3665
Crude protein value[v]	80	kg[w]	0.335	1745	585
Crude fiber value[y]	290	kg[w]		6341	
			Total value of forage in manger		4250

[z]Including losses.
[y]No positive value at date (energy + protein values exceed vetch hay price).
[x]Calculated from DM yield.
[w]Per Mg DM.
[v]Calculated after Morrison (1957).
[t]Local currency prices and US$ exchange rate.

TABLE 11.2

Comparison of roughage feeding activities in terms of land and water use — abbreviated examples. (Family farm, coastal region; 500 mm of annual rainfall, July 1980[x])

Activity		Output at the manger (per ha)					Inputs per ha				Value–cost analysis		
Yield level	Description	Fresh yield (Mg)	DM yield[z] (Mg)	Metab. energy (MgCal)	Protein (kg)	Crude fiber (kg)	Labor (h)	Irrig. water (m^3)	Total costs ($) up to: Harvest	Total costs ($) up to: Manger	Value ($) at the manger	Profit[y] ($) Per ha	Profit[y] ($) Per irrig. water
		(1)	(2)	(3)	(4)	(5)	(6)	(7)	(8)	(9)	(10)	(11)	(12)
1[w]	Alfalfa, hay	130	24.0	50400	3600	7440	143	10000	2670	3900	5060	1160	0.12
2		110	20.3	42630	3045	6290	139	10000	2670	3710	4280	570	0.06
3		85	15.7	32970	2355	4867	133	10000	2670	3470	3310	–160	–0.02
1	Alfalfa, fresh-cut	130	28.6	67210	6940	7550	250	10000	2500	3610	7030	3420	0.34
2		110	24.2	56870	5880	6380	230	10000	2500	3440	5950	2520	0.25
3		80	17.6	41360	4270	4640	200	10000	2500	3190	4330	1140	0.11
1	Spring corn on fallow, fresh-cut	100	22.0	55000	1980	5940	132	2500	1060	1910	4450	2540	1.02
2		75	16.5	41250	1180	4450	107	2500	1060	1700	3380	1680	0.67
3		50	11.0	27500	990	2970	82	2500	1060	1490	2370	880	0.35
1	Autumn corn, fresh-cut	65	13.0	31200	1170	3640	113	4000	1420	1970	2570	600	0.15
2		50	10.0	24000	900	2800	98	4000	1420	1850	1980	130	0.03
3		35	7.0	16800	630	1960	83	4000	1420	1720	1380	–340	–0.08

1	Spring corn on fallow, silage	100	21.9	52470	1740	6340	95	2500	1130	2780	4250	1470	0.59
2		80	17.4	41990	1390	5070	82	2500	1130	2460	3400	940	0.38
3		55	12.0	28890	960	3490	66	2500	1130	2050	2340	290	0.12
1	Autumn corn, silage	60	11.6	26770	930	3490	84	4000	1490	2490	2190	–300	–0.07
2		50	9.7	22300	770	2900	78	4000	1490	2330	1810	–520	–0.13
3		35	6.8	15620	540	2030	69	4000	1490	2090	1280	–810	–0.20

[z]Including losses.
[y]Profit = (value at the manger) – (total costs up to the manger).
[w]Yield level: 1, high; 2, intermediate; 3, low.
[x]Local currency prices and US $ exchange rates.

TABLE 11.3

Comparison per Mg at source of various feeding activiities — abbreviated examples. (Family farm, coastal region, 500 mm of annual rainfall, July 1980[t])

Activity					Indices for comparing activities			
					Per ton at source		Up to manger	
Yield level	Description	Fresh yield ($Mg\ ha^{-1}$)	Correct coeff.[z]	Cost in field or at purchase ($)	Total cost[y] ($)	Value ($)	Value/cost (%)	Break even ($)
		(1)	(2)	(3)	(4)	(5)	(6)	(7)
1[v]	Alfalfa, hay	130	1.00	111.2	162.5	210.8	130	20.4
2		110	0.95	105.7	177.0	210.8	119	20.4
3		85	0.90	100.1	215.5	210.8	98	20.4
1	Alfalfa, fresh-cut	130	1.00	19.2	27.8	54.2	194	45.5
2		110	0.95	21.6	30.1	54.2	179	45.5
3		85	0.90	28.6	36.7	54.2	147	45.5
1	Spring corn on fallow, fresh-cut	100	1.00	10.5	19.1	45.0	236	36.5
2		75	0.95	13.3	22.0	45.0	205	36.5
3		50	0.90	19.0	27.5	45.0	163	36.5
1	Autumn corn, fresh-cut	65	1.00	21.1	30.4	39.5	130	31.0
2		50	0.95	27.0	35.6	39.5	111	31.0
3		35	0.90	36.6	45.1	39.5	88	31.0

1	Spring corn on fallow, silage	100	1.00	11.3	27.8	42.5	153	26.0
2		80	0.95	13.4	30.0	42.5	142	26.0
3		55	0.90	18.5	35.2	42.5	121	26.0
1	Autumn corn, silage	60	1.00	24.6	41.5	36.4	90	19.7
2		50	0.95	28.3	45.1	96.4	81	19.7
3		35	0.90	38.3	55.5	36.4	81	19.7
1	Vetch hay, purchased			111.0	121.0	137	113	120
1	Cotton seed			192.0	212.0	204	96	178
1	Orange peel, fresh			11.7	16.4	284	133	23.6
1	Wheat bran			152.0	167.0	202	121	184

[z]For costs up to harvest with lower inputs as the cause of lower yield.
[y]Including harvest, transportation, storage and feeding.
[x](Energy units × value per unit) + (CP units × value per unit); values derived from prices of grain and soybean cakes, and contents of energy and protein.
[w]The price to be paid per Mg at source (in the field or at the high market so that the total cost up to the manger will equal the value at the manger.
[v]Yield level: 1, high; 2, intermediate; 3, low.
[t]Local currency prices and US $ exchange rate.

and that yields decrease because of incorrect timing, method of application, etc., and not because of the lack of inputs. Even after applying the correction coefficient it can be seen that yield level was the deciding factor in profitability of forage technologies on the farm.

The cost of home-grown forage or of purchased feed (column 3) enables a comparison of two types of feed sources: the parameter value/cost in percent (column 6) compares technologies among themselves and with the basic alternative of feeding soycake and grain. The break-even price per Mg at the source (column 7), shows how much to pay for one Mg of purchased feed, taking into account the energy, protein and fiber (in comparison with column 3) costs at the source, without transport or treatment costs, losses etc.

In Table 11.3 output data at the manger per hectare (columns 2–6) are listed for three yield levels (as in Table 11.2). The profit per hectare (column 11) is the most important parameter for the comparison among activities themselves and with activities for cash crops in order to answer the question whether allotting land and water for the production of forage crops is more profitable than producing cash crops and purchasing hay.

The calculation shows that green-chopped forage corn and alfalfa have a clear economic advantage compared with conserving them as silage or hay. The difference in the more profitable green-chopping is caused mainly by (i) physical losses during conservation and storage (and lower yield of alfalfa hay in the field), (ii) greater expenditure for conservation (processing and capital outlays), and (iii) losses in quality (lower concentration of energy and protein in the DM of conserved than in fresh forage).

Producing forage corn in the spring (the optimum growing period) was found to be more profitable than in the summer, when yields are lower and pest incidence is greater. Thus, it is more profitable to grow corn in the spring and conserve it as silage for summer and winter feeding. Ensiling corn silage in autumn is certainly not profitable when other forages are available.

Changes in technological procedures in input and in output values may, with time, change the economic efficiency of forage production activities. In Table 11.4 examples are given of changes in some indices with time, showing the influence on the output and economic efficiency (value/costs ratio) for alfalfa and silage corn. The sharp increase in the consumer price index represents inflation, which reached hundreds of percentage points in certain years; values per unit of energy and of protein, are calculated according to Morrison's equation, at given dates (based on soybean cakes and grain); output and input indices for each activity — from Table 11.3 of the normative calculations for the respective dates (base date: July 1980 = 100).

TABLE 11.4
Output-to-input ratios at different dates

Date	CPI	Monetary value (Israeli currency)		Value/cost ratio (%)		Input and output indices			
		Protein[z]	Energy[z]	Alfalfa	Corn	Alfalfa		Corn	
						Input	Output	Input	Output
(1)	(2)	(3)	(4)	(5)	(6)	(7)	(8)	(9)	(10)
July 1980	91	0.35	1.67	195	250	100	100	100	100
Oct. 1980	118	0.42	2.10	180	281	134	124	125	153
Dec. 1982	679	1.58	15.3	199	203	596	610	735	593
June 1983	1022	2.28	20.4	185	190	891	846	1107	844
Dec. 1983	1974	6.62	49.5	248	179	1774	2264	3277	2353

[z]Calculated after Morrison (1957).
CPI: consumer price index.

Table 11.4 shows that, during 1980–1983, there were differences in the price relationships between inputs and outputs (no changes occurred in the physical inputs and outputs for alfalfa and corn during this period). The price relationship between protein and energy was 4.8 in July 1980; in 1982 the increase in the price of protein was double that for the price of energy and reached almost 10; in December 1983 the ratio decreased to 7.5. The input prices also changed and for silage corn (because of higher fertilizer costs) they were higher than for alfalfa, so that value/cost ratio of corn decreased and that of alfalfa increased.

11.3 APPLICATIONS

Tables such as 11.2 and 11.3 can be used with the approach of the planting season for selecting efficient production technologies in terms of land and water use, in the light of expected and relevant crop yield levels for each of the technologies. Users whose conditions differ greatly from the normative values (Table 11.2) can adjust the calculations accordingly. Extension workers who used the computerized calculations shown above found that the corrections for other regions usually did not change much the overall classification of the activities.

Collecting the many necessary bits of information (type of inputs, quantities, prices and values of outputs) and making wide calculations are often a heavy burden for the individual user. The centralized compilation of relevant and up-to-date information by extension workers and economists can greatly facilitate the use of computerized calculations to be made on personal computers by interested growers. The spread of personal computers enables the combination of the advantages of a centralized system providing core data of information, with the particular required up-to-date details compiled on the farm itself. Programs are available to make the proper calculations. Prices change quickly, especially in times of inflation. Updated computerized brochures are prepared when deemed necessary and forwarded to interested parties.

Programs similar to those described above have been also developed for cash crops and for vegetables (Viesel et al., 1984a,b) and are regularly updated. These programs facilitate the calculation of the potential profitability of different forage production technologies in comparison with the cultivation of cash crops, in terms of water and land use. In this way the problem can be avoided of two overlapping cycles competing with on farm-grown forage crops (Fig. 11.1), i.e., with purchased feed as part of the feed ration, and with cash crops competing for water and land.

Comparative tables can also be used as guides in establishing research goals and in determining economic priorities.

11.4 CONCLUSIONS

The economic evaluation of forage production on a mixed farm under irrigation is complicated, while feeding is a major factor in animal husbandry. Many options of forage crops to be grown on the farm, their timing and form of utilization, are available, differing in cost and in their alimentary value. Other feeding alternatives exist, such as purchased feeds and by-products, and other alternatives for land and water use, like cash crops.

The contribution of farm-grown forage to the profit of the entire farm cannot be measured directly (income-cost) as it has no market price. Linear programming can provide the optimal combination of forage crops, purchased feeds and cash crops to maximize the profit of the whole farm. However, a very large matrix and much detailed technical information are needed. A collaborative effort of research, extension and farmers, and the proper use of both central and personal computers, might yield a practical and useful system.

A simpler solution (using a technique suggested by Morrison, 1957) was developed in Israel. Based on three marketable feeds, the three unknown values (i.e., market values of energy, protein and fiber for a certain date) are determined by equations, where the cost of grain, of soy-cakes and of vetch hay, equals the sums of its energy, protein and fiber contents, each multiplied by its unit value.

All forage options and purchased feed are compared on a common basis, namely, their economic value and cost up to the manger. The parameter (value/cost) per ton classifies all feeds from the point of view of animal production. The parameter (value/cost) per hectare classifies all forage and cash crops as to their economic efficiency in utilizing land and water.

Production costs and economic feed values fluctuate with time. Updated editions of these tables, issued periodically by the extension service, enabled farmers to improve their decision-making even in periods of high inflation and big changes in the relations among energy, protein and fiber economic values.

REFERENCES

Goldman, A., Goldovski, J., Zur, Y., Ben Asher, A., Baruch, M. Dar, Y. and Viesel, L., 1981. Computerized calculations for feed and cash crops. Second set: fresh cut, silage and hay crops, and purchased feeds (8.3.81 prices). Min. Agr., Ext. Serv., Tel Aviv. Updated and revised editions, 66 pp. (in Hebrew).

Morrison, F.B. 1957. Feeds and Feeding, 22nd ed., Morrison Publ. Co., Ithaca, NY. 1165 pp.

Viesel, L., Goldman, A., Goldovski, J. and Zur, Y. 1984a. Computerized calculations for field cash crops (8.3.81 prices). Min. Agr. Ext. Serv., Tel Aviv and Agr. Res. Org., Bet Dagan, Updated and revised editions, 78 pp. (in Hebrew).

Viesel, L., Goldman, A., Goldovski, J. and Zur, Y. 1984b. Computerized calculations for vegetable crops (27.5.81 prices). Min. Agr. Ext. Serv., Tel Aviv and Agr. Res. Org., Bet Dagan, Updated and revised editions, 40 pp. (in Hebrew).

COMMON AND SCIENTIFIC NAMES OF FORAGES

FORAGES

Agropyron elongatum	Wheat grass, tall
Alfalfa, lucerne	*Medicago sativa*
Alopecurus pratensis	Meadow foxtail
Astragalus cicer	Cicer milkvetch
Avena sativa	Oats
Barley	*Hordeum vulgare*
Bermuda grass, couch	*Cynodon dactylon*
Berseem, Egyptian clover	*Trifolium alexandrinum*
Beta vulgaris	Fodder, sugar beet
Blue panic	*Panicum antidotale*
Brachiaria mutica	Para grass
Bromus inermis	Bromegrass, smooth
Bromus riparius	Bromegrass, meadow
Bromegrass, smooth	*Bromus inermis*
Bromegrass, meadow	*Bromus riparius*
Buffel grass	*Cenchrus ciliaris*
Cenchrus ciliaris	Buffel grass
Centrosema ciliaris	Centro
Centro	*Centrosema ciliaris*
Chloris gayana	Rhodes grass
Cicer milkvetch	*Astragalus cicer*
Clover, alsike	*Trifolium hybridum*
Clover, white, Ladino	*Trifolium repens*
Clover, berseem, Egyptian	*Trifolium alexandrinum*
Clover, Persian	*Trifolium resupinatum*
Clover, red	*Trifolium pratense*
Clover, strawberry	*Trifolium fragiferum*

Clover, subterranean	*Trifolium subterraneum*
Cocksfoot	*see* Orchard grass
Corn, maize	*Zea mays*
Cowpea	*Vigna unguiculata*
Cynodon dactylon	Bermuda grass, couch
Dactylis glomerata	Orchard grass, cocksfoot
Dallis grass	*see* Paspalum
Digitaria decumbens	Pangola grass
Eragrostis spp.	Love grass
Fescue, tall	*Festuca arundinacea*
Festuca arundinacea	Fescue, tall
Fodder, sugar beet	*Beta vulgaris*
Green panic	*Panicum maximum* var. *T. trichoglume*
Hybrid pennisetum	*Pennisetum purpureum* × *P. americanum*
Hordeum vulgare	Barley
Kentucky bluegrass	*Poa pratensis*
Klein grass	*Panicum coloratum*
Lolium perenne	Perennial ryegrass
Lolium multiflorum	Annual, Italian ryegrass
Lolium rigidum	Wimmera ryegrass
Lotus corniculatus	Trefoil, birdsfoot
Lotus oliginosus	Trefoil, big
Love grass	*Eragrostis* spp.
Meadow foxtail	*Alopecurus pratensis*
Medicago spp.	Medic

Medic	*Medicago* spp.
Medicago sativa	Alfalfa, lucerne
Oats	*Avena sativa*
Orchard grass, cocksfoot	*Dactylis glomerata*
Pangola grass	*Digitaria decumbens*
Panicum maximum var. *trichoglume*	Green panic
Panicum antidotale	Blue panic
Panicum coloratum	Klein grass
Para grass	*Brachiaria mutica*
Paspalum, Dallis grass	*Paspalum dilatatum*
Pennisetum purpureum × *P. americanum*	Hybrid pennisetum
Phalaris, Harding grass	*Phalaris aquatica*
Phleum pratense	Timothy
Poa pratensis	Kentucky bluegrass
Rhodes grass	*Chloris gayana*
Ryegrass, annual, Italian	*Lolium multiflorum*
Rye	*Secale cereale*
Ryegrass, perennial	*Lolium perenne*
Sainfoin	*Onobrychis viciifolia*
Secale cereale	Rye
Smooth brome	*Bromus inermis*
Sorghum sudanense	Sudan grass
Sorghum bicolor	Sorghum
Sorghum	*Sorghum bicolor*
Stylo	*Stylosanthes* spp.
Stylosanthes spp.	Stylo
Sudan grass	*Sorghum sudanensis*
Timothy	*Phleum pratense*
Trefoil, birdsfoot	*Lotus corniculatus*

Trefoil, big	*Lotus oliginosus*
Trifolium alexandrinum	Berseem, Egyptian clover
Trifolium hybridum	Alsike clover
Trifolium repens	White, Ladino clover
Trifolium resupinatum	Persian clover
Trifolium subterraneum	Subterraneum clover
Trifolium fragiferum	Strawberry clover
Triticum aestivum	Wheat
Vetch	*Vicia sativa*
Vigna sinensis	Cowpea
Wheat	*Triticum aestivum*
Wheat grass, tall	*Agropyron elongatum*
Wimmera ryegrass	*Lolium rigidum*
Zea mays	Corn, maize

OTHER PLANTS

Cuscuta spp.	Dodder

Subject Index